Kangaroos: their ecology and management in the sheep rangelands of Australia

Cambridge Studies in Applied Ecology and Resource Management

The rationale underlying much recent ecological research has been the necessity to understand the dynamics of species and ecosystems in order to predict and minimise the possible consequences of human activities. As the social and economic pressures for development rise, such studies become increasingly relevant, and ecological considerations have come to play a more important role in the management of natural resources. The objective of this series is to demonstrate how ecological research should be applied in the formation of rational management programmes for natural resources, particularly where social, economic or conservation issues are involved. The subject matter will range from single species where conservation or commercial considerations are important to whole ecosystems where massive perturbations like hydro-electric schemes or changes in land-use are proposed. The prime criterion for inclusion will be the relevance of the ecological research to elucidate specific, clearly defined management problems, particularly where development programmes generate problems of incompatibility between conservation and commercial interests.

KANGAROOS

their ecology and management
in the sheep rangelands
of Australia

Edited by
Graeme Caughley
Division of Wildlife and Rangelands Research, CSIRO, Australia
Neil Shepherd
New South Wales National Parks and Wildlife Service, Australia
Jeff Short
Division of Wildlife and Rangelands Research, CSIRO, Australia

A joint project of
New South Wales National Parks and Wildlife Service and
CSIRO Division of Wildlife and Rangelands Research

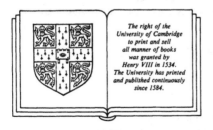

The right of the
University of Cambridge
to print and sell
all manner of books
was granted by
Henry VIII in 1534.
The University has printed
and published continuously
since 1584.

CAMBRIDGE UNIVERSITY PRESS

Cambridge

London   New York   New Rochelle

Melbourne   Sydney

CAMBRIDGE UNIVERSITY PRESS
Cambridge, New York, Melbourne, Madrid, Cape Town, Singapore,
São Paulo, Delhi, Dubai, Tokyo

Cambridge University Press
The Edinburgh Building, Cambridge CB2 8RU, UK

Published in the United States of America by Cambridge University Press, New York

www.cambridge.org
Information on this title: www.cambridge.org/9780521123402

© Cambridge University Press 1987

First published 1987
This digitally printed version 2009

A catalogue record for this publication is available from the British Library

Library of Congress Cataloguing in Publication data
Kangaroos, their ecology and management in the sheep rangelands.
    (Cambridge studies in applied ecology and resource management)
    'A joint project of New South Wales National Parks and Wildlife Service and
CSIRO Division of Wildlife and Rangelands Research.'
        1. Kangaroos – Ecology.
        2. Wildlife management – Australia – New South Wales.
        3. Range management – Australia – New South Wales.
        4. Range ecology – Australia – New South Wales.
        5. Mammals – Ecology.
        6. Mammals – Australia – New South Wales – Ecology.
        I. Caughley, Graeme.
        II. Shepherd, Neil.
        III. Short, Jeff.
        IV. New South Wales National Parks and Wildlife Service.
        V. Commonwealth Scientific and Industrial Research Organization (Australia).
            Division of Wildlife and Rangelands Research.
        VI. Series.
QL737.M35K38   1987   639.9'792   86-23314

ISBN 978-0-521-30344-6 Hardback
ISBN 978-0-521-12340-2 Paperback

CONTENTS

[A] New South Wales National Parks and Wildlife Service, Box N189,
Grosvenor Street P.O., Sydney, N.S.W., 2000, Australia.

[B] Division of Wildlife and Rangelands Research, CSIRO, P.O. Box 84,
Lyneham, A.C.T., 2602, Australia.

PREFACE

The research project that forms the basis of this book was commenced within the New South Wales National Parks and Wildlife Service in 1977. It was designed primarily to examine the relationship between high kangaroo densities and vegetation in an arid-zone national park (Kinchega National Park). By examining kangaroo dynamics in relation to weather and pasture availability, and studying kangaroo movement, diet, and techniques for monitoring population change and well-being, it was also hoped to provide information relevant to management of kangaroos on rangelands. Most of the research on the Park was duplicated on the adjoining grazing lease (Tandou).

The years 1977 to 1980 saw the development of techniques for handling the peculiarities of rangeland vegetation, weather and herbivores. In late 1979 CSIRO Division of Wildlife Research (now the CSIRO Division of Wildlife and Rangelands Research) was invited to participate. The invitation was accepted and a formal agreement was drawn up in August 1980 to conduct a joint project for five years. The project was then modified and expanded, proceeding to the end of the contract period in November 1985. This book deals largely with the period of the joint study.

The full scope of this project is not covered in this volume. Two major omissions are the effect of grazing by rabbits and the effect of insects on the vegetation. Neither investigation is sufficiently advanced for inclusion. However, results available at the time of writing suggest that neither rabbits or insects exerted major effects over the period of this study. Authorships of the various chapters of this book reflect responsibility for a segment of the project, but most of the authors contributed to the field work and writing of chapters that do not bear their names. Many other people contributed to the process of turning ideas into completed scientific work. An attempt is made in the succeeding paragraphs to acknowledge their assistance.

First we acknowledge the assistance of the 400 or so volunteers, mostly undergraduate students, without whose toil this project could never have functioned. Secondly, but of no lesser importance, are the management staff of the Western Region of the New South Wales National Parks and Wildlife Service; in particular John and Kath Eveleigh, Peter Morris, Peter Evans, Barrie Booth, John McDonald and the Kinchega Work Staff, and Greg Tucker. Bob Smith, manager of Tandou, allowed us the run of Tandou. Ron Lukoschek, its livestock manager, was unstinting with information and assistance. Within the Service the following people deserve special mention: Don Johnstone, Jack Giles, Keith Mullette and John Whitehouse for initial approvals and for continued support; Graeme Dalitz, Richard Byrne, Graham Fleming, Richard Howell, and Ross Joy for providing aircraft and piloting them; Michael O'Connell for statistical advice; Judy Caughley for critical reading; and Bryan Spurrs for administrative support. Jack Giles, John Taylor (University of New England), and Bob Harden were important sources of advice and experience in the early stages of the project. Many other people helped in the field, laboratory, or computer room.

Within CSIRO we thank Paul Magi and Albert Williams for administrative support, Dean Graetz, Henry Nix and John Calaby for discussion and critical reading, Chris Davey for computing assistance, Jo Smith, Bevan Brown and David Grice for assistance in the field, the last also for analysis and for acting as text manager, Edric Slater and Graham Chapman for additional photography, and Frank Knight for drafting. Finally we are particulary grateful to Murray Ellis of the Service who produced almost all the figures and kept us organised.

The opinions expressed in this book are those of the authors, not necessarily those of the organisations to which they belong.

Graeme Caughley
Neil Shepherd
Jeff Short

Inhabits the western side of *New Holland,* and has as yet been discovered in no other part of the world. It lurks among the grass: feeds on vegetables: goes entirely on its hind legs, making use of the fore-feet only for digging, or bringing its food to its mouth. The dung is like that of a deer. It is very timid: at the sight of man flies from them by amazing leaps, springing over bushes seven or eight feet high; and going progressively from rock to rock. It carries its tail quite at right angles with its body when in motion; and when it alights often looks back: it is much too swift for greyhounds: it is very good eating.

It is called by the natives, *Kanguru.*

T. Pennant, *History of Quadrupeds.* London, 1781.

I come from the Northern plains
 Where the girls and grass are scanty;
Where the creeks run dry or ten feet high
 And it's either drought or plenty.

The Overlander (Traditional Ballad).

THE STUDY AREA IN CONTEXT

Fig. 1.1. Map of New South Wales showing the location of the study area and other localities referred to in the text. The inset shows the states and territories of Australia. Clockwise from the left they are Western Australia, Northern Territory, Queensland, New South Wales, Victoria, Tasmania and South Australia.

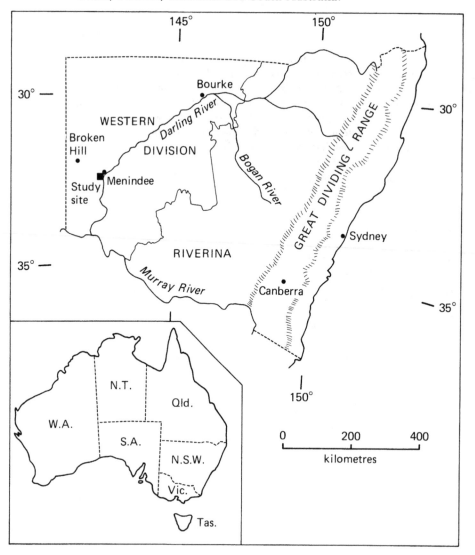

1

Introduction to the sheep rangelands

GRAEME CAUGHLEY

1.1   Context and questions

In this book we examine the sheep rangelands of Australia, not as they were but as they are today. We focus particularly on the chenopod shrublands, a plant association that covers 5-10% of the arid and semi-arid areas of Australia (Goodall, 1979) and which occurs mainly in the winter rainfall zone (150-500 mm per year) of southern Australia (Oxley, 1979).

Our main region of study is the Menindee District (Fig. 1.1) of western New South Wales. Within this we studied two areas, the first being Kinchega National Park (440 km²). Here we examined the structure and dynamics of a grazing system comprising a vegetation dominated by perennial chenopods and annual grasses and forbs, grazed and browsed by a suite of herbivores dominated by the red kangaroo and the western grey kangaroo. The second study area is Tandou sheep station (810 km², of which we measured the vegetation on 200 km² and the kangaroos on about 400 km²) which abuts Kinchega to the south. There we studied much the same system as on Kinchega, but with sheep added to it.

Rainfall averages 236 mm per year but varies prodigiously between years (see Chapters 2 and 3). The weather is the driving variable of this arid-zone system, largely determining the biomass of the pasture layer month by month (see Chapter 4) and through its influence on that layer determining what the herbivores eat at any given time (see Chapter 5), how much they eat (see Chapter 6), and how far they have to move to get it (see Chapter 7). Those constraints in turn determine the condition and reproductive success of the herbivores (see Chapter 9) and whether they increase or decrease in numbers (see Chapter 8). This sequence of consequences loops back to the vegetation through the effect of grazing and browsing on plant composition and plant biomass.

We were lucky with the weather during this study. It fortuitously exhibited its full repertoire: times of plenty, heavy drought, and recovery to plenitude. The system was tossed around by these perturbations, allowing us to calculate rates of change of plants and animals at levels of pasture biomass ranging between extremes separated by two orders of magnitude. We use these data to probe the outer limits of the system's dynamics within the national park and to determine, within those limits, how a change in one component affected the others. Concomitantly, we studied in lesser detail the ecological system on sheep stations outside the national park where the ecology of the park, complicated enough as it is, is complicated further by the superimposition of sheep.

Finally, having achieved modest insight into how the ecological system operates, with and without sheep, we apply this knowledge to the practical questions of how kangaroos may best be managed in national parks where one set of perceived goals obtains, and how they might be managed on rangelands where the perceived goals are very different. We offer options rather than prescriptions because different groups of people cleave to different values and so have different goals.

1.2   Background

1.21   Prehistory

People discovered Australia at least thirty thousand years ago. The dating of the arrival is messy but we know that an adult male was buried with due ceremony on the shore of Lake Mungo in western New South Wales sometime in the period 28,000 to 30,000 years ago (Bowler & Thorne, 1976). Clearly the first arrivals predated that, but by how much we do not know. Guesses of 10,000 and 20,000 years are common (White & O'Connell, 1982), and it has even been suggested from the incidence of charcoal in sedimentary profiles that 50,000 or more years could be added to the Lake Mungo date (Singh, Kershaw & Clark, 1981). My guess is modest — no more than 4,000 years.

The people came in from the north-west, perhaps first into New Guinea which at that time was linked to Australia. The Pleistocene sea level then was 100 m lower than now. The continental shelf was dry land. These people found a land less arid than today, one stocked with many species of large mammals. About 30,000 years ago Australia had a megafauna (species over 50 kg) of 13 marsupial genera, a genus of gigantic monitor lizards, and a genus of heavy flightless birds. All species of 87% of those genera had become extinct by some time before 18,000 years ago.

In the now semi-arid and arid regions of eastern Australia the greatest decline of the megafauna occurred before 25,000 years ago, but species extinct in that region may have lingered on for up to 6,000 years along the southern periphery of the continent (Hope, 1982). The most economic explanation for these extinctions and for those in other parts of the world is that the animals were hunted out, but that is widely rejected in favour of the 'climatic hypothesis'. It requires that modest changes in climate in the late Pleistocene and Holocene caused bursts of extinction in different parts of the world at different times even though the previous climatic fluctuations of the Pleistocene did not raise the frequency of extinction much above its background rate. There are large problems with this hypothesis.

In contrast, all data are consistent with overkill. The extinctions were biased strongly towards big mammals, those whose intrinsic rates of increase are the lowest and consequently those whose dynamic reaction to lowered density is the weakest. The way out of that bind is by dwarfing, and that certainly occurred (Hope, 1982). Although a special case for the climatic hypothesis can be made for any one area, it falls apart when the pattern of extinctions is looked at across space and time. The one consistent correlation is with the rise and spread of hunting cultures. It is easy to show by computer simulation (e.g. Martin, 1973) that human populations at low density are entirely capable of generating the hunting pressure needed to slide a population of big mammals to extinction. All one needs is a few spears, a few mates and a few centuries. White & O'Connell (1982) missed the point when criticising this model of the spread of primitive people and their possible effect on the megafauna of the Americas: 'it seems highly unlikely that they would have been able to press on in monomaniacal pursuit of these animals throughout the entire continent . . . Would tradition have been sufficient to keep a majority of the population moving forward scores of kilometres a year?' In fact the human dispersal front need have moved no more rapidly than 11 km per year (Caughley, 1977) to be consistent with the data, and the 'majority of the population' would not have moved at all.

After the megafaunal collapse and before the advent of pastoralism, Australia was to receive another ecological jolt, the dingo. The earliest reliable dates of its occurrence on the Australian mainland (it did not reach either New Guinea or Tasmania) lie between BC 1,500 and BC 1,000. Its relationships are with central and southern Asian domestic dogs (Gollan, 1984). How it got to Australia is not known but I would hazard a guess that the Aborigines had nothing to do with its introduction. I detect instead the fingerprint of a seafaring people and would nominate the group that gave rise to the Macassans of the Celebes (the group that colonised Madagascar) as the most likely. What effect the dingo had can only be conjectured but the disappearance of the marsupial tiger *Thylacinus* from the mainland at about the same time (last firm date BC 1,000) may be more than coincidence.

1.22   *The advent of pastoralism*

Sheep and cattle were herded onto the inland plains of Australia soon after that region was first traversed by European explorers. Frith (1973) summarised the subsequent events in the Riverina of New South Wales: the Lachlan, Murrumbidgee and Murray Rivers system. Sturt explored the region in 1829 and the pastoralists followed him in 1835, the major settlement being in 1839. Thanks to Krefft (1866) of the Blandowski Expedition we know that in 1857 the area contained 20 species of marsupials and five species of rodents. Today it contains only six species of those marsupials and one of the original five rodents. The patterns of extinction differed in kind from those of the late Pleistocene. The wallabies, bandicoots, dasyures and rodents that disappeared were all small animals that lived on the ground and were dependent on a thick layer of herbs and grasses for cover. The main cause of the extinctions seems to have been modification of the habitat by the sheep and cattle, although rabbits (which entered the area in 1881) and eutherian predators also may have played a part. In sharp contrast, the more closely settled eastern seaboard and the coastal ranges lost not one mammalian species (Calaby, 1966). The fauna of the sheep rangelands is examined in greater detail in Chapter 2.

Against those losses must be balanced the gains. Pastoralism formed the base of an emerging nation which rode into the twentieth century on the sheep's back. The losses may be regretted but they were largely inevitable. The injection of an exotic and husbanded herbivore into an ecosystem whose components had been selected to cope with quite different forces broke the system down to a new and more simple structure. The changes wrought by grazing were augmented by management designed to change the environment further in favour of livestock. Edaphic forest was converted to induced grassland. Wildlife that killed stock were eliminated. Indigenous species that competed with stock were controlled. Dispersal was modified by fencing and artificial watering points were created. The system was changed beyond recognition.

Figure 1.2 shows the trend of sheep numbers in western New South Wales from soon after establishment to the present. It followed the eruptive fluctuation typical of a herbivore colonising a new environment (Riney, 1964; Caughley, 1970). The upswing was a response to a capital supply of food containing many species of plants that could not tolerate the grazing and browsing of the alien herbivores. These species declined in density under grazing and, when the system sorted itself out, remained in the vegetation only as remnants. The herbivore peak represents a population that had climbed to high density up the ladder of edible biomass provided temporarily by those declining species. The crash is a consequence of the population being stranded at high density in the face of a now greatly reduced standing crop of edible vegetation. The subsequent trend of lower density with little long-term time trend but much short-term fluctuation in response to variable weather, marks the post-colonisation period where the vegetation and the sheep have reached an uneasy truce. A new biological regime has been created. That summary is broadly consistent with the conclusions of Williams & Oxley (1979), with the difference that they ascribe more importance than I do to the influence of high rainfall and drought on the peak and crash of sheep numbers at the turn of the century.

Figure 1.4 gives a modelled outcome of the process as it would probably occur in the absence of year-to-year variation in weather. It assumes a plant population growing logistically in the absence of grazing, and a population of herbivores whose rate of increase per head is a monotonic and asymptotic function of edible plant biomass per unit of area.

Fig. 1.2. Trend of sheep numbers in the rangelands of New South Wales. Data from Butlin (1962) for the 'Western Division' 1860-1920 and from Australian Bureau of Statistics for the 'Western Plains' 1921-1984. The two regions are essentially congruent but with minor differences in their eastern boundaries.

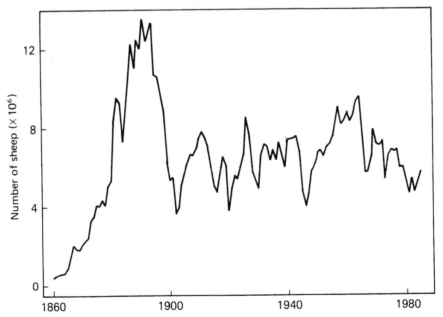

Fig. 1.3. The face of western New South Wales since European settlement: sheep, kangaroos and waterholes.

Fig. 1.4. Modelled trend of pasture biomass and animal numbers during an ungulate eruption (from Caughley, 1976b).

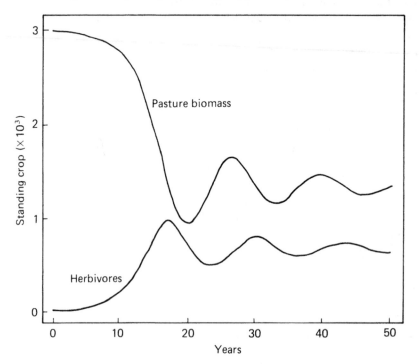

1.23   *Sheep and kangaroos*

Figure 1.5 is a map of the land-use zones of Australia. The sheep rangelands form a broken crescent, one half in the east and the other in the west. On its seaward side is the sheep-grain zone of arable farming and intensive sheep production. Inland of it is the zone of extensive cattle grazing.

Figure 1.6 maps the distribution of sheep in Australia. These are concentrated in the sheep rangelands and the sheep-grain belt.

Figure 1.7 shows how the red kangaroo *Macropus rufus* is distributed throughout Australia, the highest densities being in the sheep rangelands of the east. Figure 1.8 provides the analogous map for the eastern grey kangaroo *Macropus giganteus*. The distribution is truncated on this map at the western slopes of the Dividing Range but the range of this species continues to the coast. It occurs at highest densities in the sheep rangelands of eastern

Fig. 1.5. Land use in Australia, and the Menindee study site.

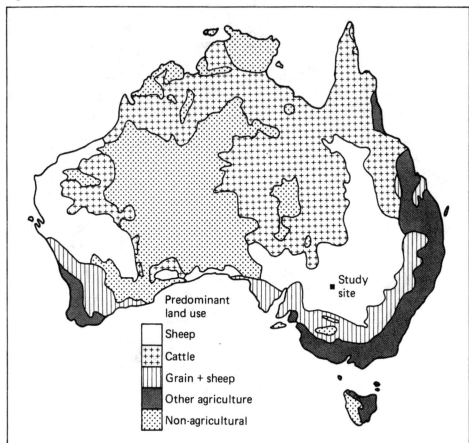

Study site

Predominant land use

Sheep

Cattle

Grain + sheep

Other agriculture

Non-agricultural

Australia, particularly in the east of that zone on both sides of the border between Queensland and New South Wales.

Figure 1.10 shows the spatial variation in density of the western grey kangaroo *Macropus fuliginosus*. (In terms of its distribution it is misnamed, extending as it does across the south of the continent in areas of winter rainfall and into dry areas where summer and winter rainfalls are similar. 'Southern grey kangaroo' has been suggested as a more appropriate name (Caughley, Grigg & Short, 1983) but here we retain the more familiar 'western grey'.)

These three species of kangaroo reach their highest densities in the sheep rangelands (Table 1.1) and it is mainly in that land-use zone that they are seen as conflicting with agriculture. (In Australia 'agriculture' tends to be reserved for arable farming with animal production being termed 'pastoralism'

Fig. 1.6. Distribution of sheep in Australia in 1977. Data from Atlas of Australian Resources, 3rd series, 1979. Note that most sheep are outside the rangelands.

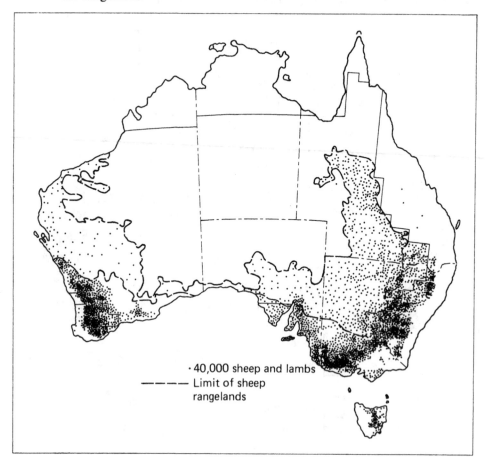

· 40,000 sheep and lambs
— — — Limit of sheep rangelands

or simply 'grazing'. That etymological dichotomy leads to a further complication whereby 'agriculture' is what is done by 'farmers' whereas animal production is the domain of 'graziers', unless they produce milk which makes them 'farmers' again. We will not allow these sociological distinctions to get in our way.) No state other than Queensland allows hunting of these three kangaroo species for sport. Their control is instead entrusted to an industry which harvests them under licence and releases the meat and skins to domestic and overseas markets. Additionally, land occupiers may be issued 'damage mitigation' permits to shoot kangaroos but the animals taken under these must not be sold.

Fig. 1.7. Density and distribution of the red kangaroo, from aerial surveys 1980-1982.

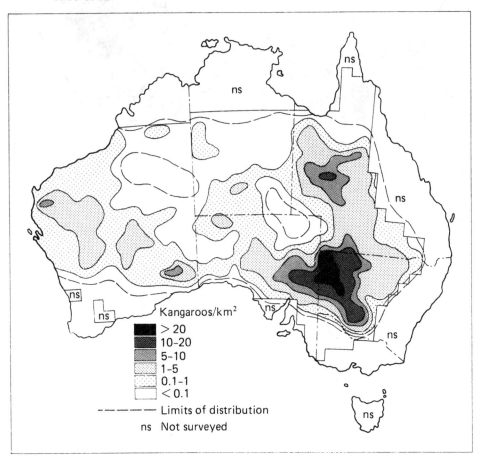

Fig. 1.8. Density and distribution of the eastern grey kangaroo over the inland plains portion of its range, from aerial surveys 1980-82. Densities are conservative.

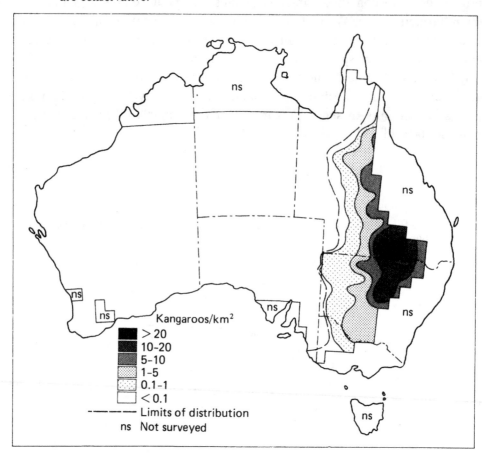

Fig. 1.9. The 1.8 m high 'macropod deterrent' fence that proved a leaky boundary for Kinchega National Park. Most sheep fences are built of strands of plain wire, are half as tall and do not act as a barrier to kangaroos.

Fig. 1.10. Density and distribution of the western grey kangaroo, from aerial surveys 1980-82. Densities are conservative.

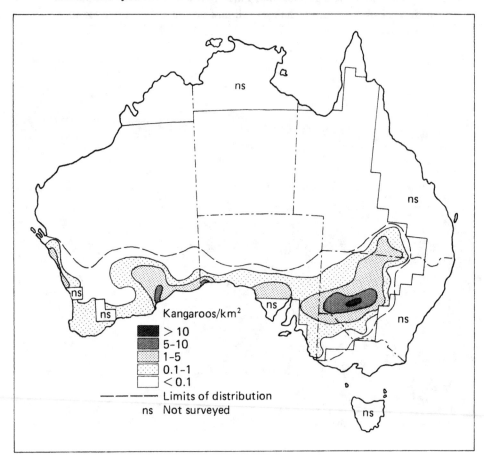

Table 1.1. *Estimates in millions of the number of sheep, red kangaroos,
eastern grey kangaroos and western grey kangaroos in the mainland
states of Australia in 1981, the states being divided into the zone of
'rangelands' used for extensive sheep grazing and the 'other' zones used
for such purposes as wheat farming, forestry, and extensive cattle
grazing. A dash indicates either that a species is not present in a zone
(e.g. no eastern greys in Western Australia) or that the land-use category
does not exist in that state (e.g. almost no 'rangeland' in the sense used
here in Victoria). The) symbol before an estimate indicates that only
part of the zone was surveyed. Estimates for grey kangaroos are conservative. Kangaroo numbers per zone are a breakdown of the state estimates
in Caughley, Grigg and Short (1983). Sheep numbers are extracted from
statistics published by the government of each state.*

| Species | Zone | Australian Mainland States | | | | | |
		NSW	VIC	QLD	SA	WA	Total
Sheep	Rangeland	5.5	—	9.8	3.1	2.1	20.5
	Other	43.2	25.3	2.6	13.6	28.1	112.8
	Total	48.7	25.3	12.4	16.7	30.2	133.3
Red	Rangeland	3.5	—	1.3	1.2	0.6	6.6
kangaroo	Other	0.3	trace	0.9	0.1	0.5	1.8
	Total	3.8	trace	2.2	1.3	1.1	8.4
Eastern	Rangeland	1.4	—	2.5	—	—	3.9
grey	Other	)0.5	)0.005	)0.5	trace	—	)1.0
kangaroo	Total	)1.9	)0.005	)3.0	trace	—	)4.9
Western	Rangeland	0.8	—	0.1	0.3	0.1	1.3
grey	Other	0.04	0.02	—	0.08	0.3	0.4
kangaroo	Total	0.84	0.02	0.1	0.38	0.4	1.7

2

The environment of the Australian sheep rangelands

GRAHAM ROBERTSON, JEFF SHORT
AND GREG WELLARD

2.1   Introduction

The sheep rangelands, a subset of the arid and semi-arid lands of Australia, form a broad semi-circular arc from northern Queensland, through New South Wales, South Australia and the southern two-thirds of Western Australia (Fig. 1.5). It covers approximately 1.7 million km² or 22% of the area of Australia. Climate determines the outer margin of the zone in the southern half of the continent: being the amount and incidence of rain sufficient to cultivate crops, particularly wheat. In southern New South Wales, South Australia and Western Australia this approximates the 260-290 mm isohyet. In southern Queensland the sheep rangelands extend into country receiving nearly 600 mm of rain per annum. The inner margin, where the sheep pastoral zone abuts either cattle rangelands or desert, is at about the 150 mm isohyet in the south and the 300 mm isohyet in the north.

This chapter examines the climate, soils, vegetation and fauna of the sheep rangelands of Australia, focusing on Kinchega National Park and a neighbouring sheep station, Tandou. These are located near the town of Menindee (elevation 61m) at approximately 32 degrees S and 142 degrees E in western New South Wales on the floodplain of the Darling River. Almost all data in succeeding chapters were collected on Kinchega and Tandou. Both were formerly part of a much larger pastoral lease and have been grazed by sheep since 1860. In 1967 Kinchega was declared a national park under the control of the New South Wales National Parks and Wildlife Service. The Park (44,000 ha) was fenced to keep kangaroos in and sheep out. Tandou comprises 81,000 ha of which 18,600 ha (the dry bed of Lake Tandou) is cropped intensively by irrigating from the anabranch of the Darling River. Tandou is approximately three times the size of the average pastoral lease in the district. Its rated stocking capacity is 0.2 sheep per hectare.

2.2   Climate

This section examines the climate of the sheep rangelands of Australia, illustrates its characteristics by reference to the long-term weather data for Menindee (a meteorological station 10 km north-east of Kinchega National Park) and examines the weather at Kinchega during this study (1981-83 inclusive).

2.21   *Climate of the sheep rangelands*

The climate of the Australian sheep rangelands is characterised by high summer temperatures, low and erratic rainfall and a level of evaporation that greatly exceeds rainfall. Figure 2.1 maps the distribution of rainfall on the Australian continent and indicates seasonal incidence. The southern half of the continent may experience widespread rain in winter with passage of

Fig. 2.1. The distribution and seasonal incidence of rainfall on the Australian continent. The isohyets are in millimetres.

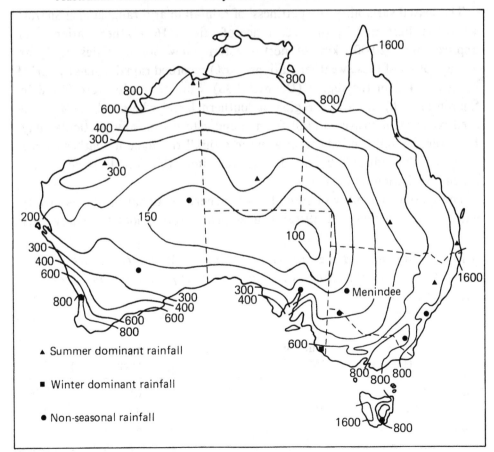

cold fronts from west to east. The tropical monsoon brings rainfall in summer to the northern half of the continent. Cyclonic rainfall (heavy rainfall accompanied by strong winds and associated with intense low pressure systems) falls in northern Australia in summer and may penetrate southern Queensland and northern New South Wales. The interior of the continent receives rainfall in summer from thunderstorms.

Menindee receives a mean annual rainfall of 236 mm distributed almost uniformly between seasons. The 100 year rainfall record is presented in Fig. 2.2. The reliability of rainfall in a given year and season is low (Table 2.1).

Figure 2.3 gives serial correlograms for annual and seasonal rainfall. Correlation between rainfalls of successive years is weak with no evidence of a cyclicity of longer duration. The correlogram of seasons seeks cylicity among periods of three months. Thus the correlation against 4 on the horizontal scale links spring with spring, summer with summer, and so on, in successive years. It is an index of seasonality and indicates that seasonality of rainfall is almost non-existent.

The spatial variability or 'spottiness' of rainfall in arid rangelands (Sharon, 1972a) is illustrated by the occurrence of rain at 16 weather stations over approximately 20,000 km² of north-western New South Wales. Only an average of six of these weather stations (40%) recorded rain during any period when rain fell in the region (Denny, 1983). Similar results were found by Sharon (1972b) in an arid region of southern Israel where only 20% of the area received intensive rainfall from thunderstorms on any particular day. However the uniform long term average rainfall throughout southern Israel indicated that rainfall balanced out over time, the thunderstorms being randomly distributed in space.

The spatial variability of rainfall on Kinchega National Park was investigated using a network of 39 raingauges placed throughout the study area.

Table 2.1. *Annual and seasonal rainfall (mm) at Menindee and its reliability. Seasonal rainfall is the mean rainfall for three months for the period 1883–1984.*

	Season				
	Summer (Dec.–Feb.)	Autumn (Mar.–May)	Winter (June–Aug.)	Spring (Sept.–Nov.)	Year
Rainfall	62	57	59	61	236
SD	59	47	34	44	107
CV (%)	94	84	60	73	47

Spatial variability was high, varying by as much as 100% over 3 km (Table 2.2).

Winter rainfall at Menindee tends to be of low to medium intensity. It results from cold fronts and thus tends to extend over wide areas. In contrast, summer rainfall comes as localised thunderstorms of high intensity and short duration.

Fig. 2.2. Rainfall at Menindee, New South Wales between 1880 and 1984.

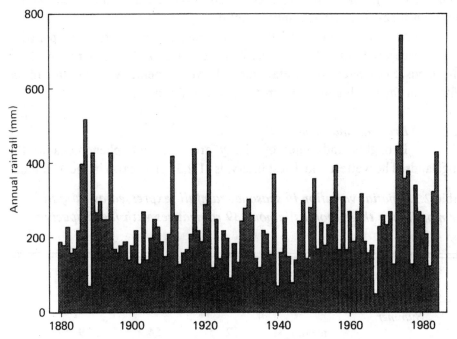

Fig. 2.3. Serial correlograms for annual and seasonal rainfall between 1885 and 1984.

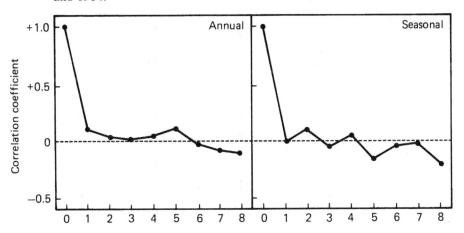

Seasonal maximum and minimum temperatures at Menindee vary more than those on the east or west coast of the continent due to the absence of the stabilising influence of the oceans on temperature fluctuations in the interior of the continent. In many inland locations across Australia including Menindee, temperatures exceed 40°C in summer. Temperatures experienced during this study (Fig. 3.2) were similar to the long term averages.

Evaporation at Broken Hill (110 km north-west of Menindee) averages 906 mm in summer, 492 mm in autumn, 244 mm in winter and 621 mm in spring (Gentilli, 1971). The yearly average evaporation of 2,263 mm at this site is about ten times its average annual rainfall.

Low and variable rainfall, high summer temperatures, and an evaporation greatly exceeding rainfall are consistent features of the Australian sheep rangelands. Seasonality of rainfall varies however, being winter-dominant or uniform in the south and summer-dominant in the north.

2.22   *Drought and plenty*

Droughts and times of plenty are recurring phenomena of the rangelands. The wettest and driest intervals (1,2,3....10 years) in the Menindee

Table 2.2. *Spatial variation in seasonal rainfall (expressed as rainfall per month over three months) among 39 raingauges with mean spacing of 3.4 km.*

		Year 1981	Year 1982	Year 1983
Season		1981	1982	1983
Summer				
	mean	12	54	9
	SD	8.6	33.9	20.4
	CV(%)	72	63	224
Autumn				
	mean	2	11	39
	SD	4.0	13.8	25.7
	CV(%)	200	131	66
Winter				
	mean	45	5	31
	SD	38.5	6.0	23.3
	CV(%)	86	112	75
Spring				
	mean	14	29	34
	SD	12.0	24.7	22.6
	CV(%)	88	85	67

rainfall record are shown in Table 2.3. The 1970s were the wettest period for the previous 100 years. The wettest year, 1974, received three times the average rainfall. The driest year, 1967, received only 22% of the mean annual rainfall. The drought of the 1940s provided the driest intervals of 2, 3, 5, 6, 7 and 9 years in the rainfall record but 1895-1904 was the driest period of ten years. The 1982-83 drought, covered by this study, was less severe at Menindee than were several others this century.

When does a period of rainfall deficiency become a drought? Definitions of drought vary with environment. Thus in Britain, an absolute drought is defined as 15 consecutive days on each of which less than one point (0.254 mm) of rain falls (Foley, 1957). However, long dry spells are a normal occurrence in Australia and the definition is clearly inadequate. The term

Table 2.3. *Drought and plenty at Menindee. The wettest and the driest periods in the 100 year rainfall record (1884–1983). W and D equal the wettest and driest year or years in the rainfall record for that interval of years.*

Years in interval	Interval	Average annual rainfall (mm)
1	W 1974	739
	D 1967	52
2	W 1973–74	591
	D 1944–45	110
3	W 1973–75	514
	D 1943–45	123
4	W 1973–76	480
	D 1964–67	146
5	W 1973–77	410
	D 1940–44	145
6	W 1972–78	399
	D 1940–45	144
7	W 1973–79	380
	D 1940–46	158
8	W 1973–80	366
	D 1922–29	168
9	W 1973–81	349
	D 1940–48	173
10	W 1971–80	333
	D 1895–1904	179

drought in Australia is used to signify 'a period of months or years during which little rain falls and the country gets burnt up, grass and water disappears, crops become worthless and sheep and cattle die' (Russell, 1896, quoted in Foley, 1957).

Recent theories on the global causes of Australian droughts link the surface temperature of the ocean off the east coast of Australia with continental rainfall deficiencies. A drop in the temperature of these waters leads to a reduction in atmospheric moisture and a resulting reduction in rainfall (Streten, 1981, 1983). The fall in temperature of the sea-surface may be linked to the El Nino-Southern Oscillation phenomena which is postulated to affect weather as far afield as India and North America (Philander, 1983).

2.3   Landforms and soils

2.31   *The sheep rangelands*

The sheep rangelands are predominantly flat. They include both erosional and depositional surfaces. Plains of low altitude ($\langle 350$ m) are the principal landform. The rangelands in Western Australia are dominated by granitic and limestone plains interspersed with areas of floodplain and wind-blown sand. Those in South Australia, New South Wales and Queensland include depositional floodplains, sandplains and dunefields, and erosional surfaces.

The soils of the sheep rangelands include calcareous earths, red and yellow earths, cracking clays, red sands, clayey sands and lithosols. Of these the first three predominate. Cracking clays and calcareous earths and sands, the common soils at Menindee, form about 15% of the soils of the sheep rangelands.

2.32   *The study area*

Kinchega and Tandou lie on the floodplain of the Darling River. They contain areas of grey clay soils that are irregularly flooded, plus higher sandplains and dunes. This spatial heterogeneity is not found in the sheep rangelands away from rivers or major watercourses. A complex mosaic of 16 landclasses is recognised (The Soil Conservation Service of New South Wales). These are grouped into three major landforms: dunefields and sandplains, floodplains, and overflow lakes (Fig. 2.4).

Dunefields and sandplains occupy approximately two-thirds of Kinchega and over one-half of Tandou. They are characterised by widely spaced longitudinal sand dunes of low relief ($\langle 10$ m) trending west-east. Sandplains are slightly undulating to flat areas between dunes or on their margins. Dunes are aeolian red sands. Soils in the swales are silty clays, solonised brown soils or clay loams. Texture-contrast soils are common. Soils of the sandplains consist of red earths with clay loam or hard-packed fine sand surfaces.

Floodplains occupy approximately one-third of Kinchega and one-half of Tandou. They are a mosaic of landforms and soil types formed from the

Fig. 2.4. The major landforms of Kinchega National Park and Tandou Property.

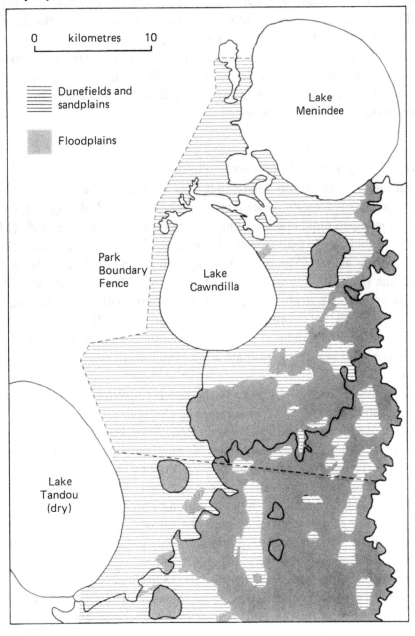

floodplain, billabongs, swamps and creeks of the Darling River. While outliers of dunes and sandplains occur throughout, the area essentially consists of flat, heavy-textured soils that are innundated on average one year in seven. The soils all have alluvial deposits of deep heavy-textured grey clay except for the higher rises; these have texture-contrast soils or deep red sands. The grey clays expand when wet and may become self sealing in heavy rain. They contract, crack deeply and become crumbly when dry. They are very deep and are particularly resistant to erosion. Soils on much of the floodplain and in the depressions and drainage tracts have gilgai microrelief.

Overflow lakes (e.g. Lakes Menindee and Cawndilla) are characterised by extensive sand dunes on their leeward sides. Dunes and sandplains grade into the overflow lakes on their windward sides. Soils on the dunes consist of both hard-packed and loose white sands.

2.4   Vegetation

2.41   *Vegetation changes since settlement*

Vegetation of the sheep rangelands has changed substantially since the introduction of domestic stock and rabbits. Postulated changes include the reduction of perennial grasses in the wetter portions of the rangelands (Moore, 1962), reduction of small trees and chenopods in the chenopod shrubland of New South Wales (Moore, 1953a, b), reduction of chenopods in South Australia (Lay, 1979; Fatchen, 1978), an increase in the incidence of woody shrubs unpalatable to stock in central New South Wales (Harrington, Oxley & Tongway, 1979), the introduction of exotic species (Dixon, 1892; Moore, 1959) and the failure of long-lived tree species to regenerate despite favourable climatic sequences for germination and establishment (Lange & Purdie, 1976).

Before the arrival of Europeans the sheep rangelands were grazed by a different suite of herbivores. There were no sheep. According to Oxley (1820), Mitchell (1839) and Sturt (1849) large kangaroos were in low numbers. The country was grazed by several species of small marsupials, each weighing less than 3 kg. These presumably had different grazing behaviour and preferences from the large kangaroos or introduced livestock. If heavy grazing did occur it was probably around waterholes and along watercourses.

That grazing regime changed with the introduction of sheep. In the few decades after settlement sheep numbers expanded rapidly reflecting the squatters' optimism in the carrying capacity of the country and their ignorance of the forthcoming droughts. In the late 1890s western New South Wales supported 14 million sheep, about twice the current rated carrying capacity. One property near Broken Hill carried 90,000 sheep on 160,000 ha or more than 1 sheep to 2 ha on country which today carries 1 sheep to 6 ha (Condon, 1982). These sheep were watered at an average of 5,000 sheep per watering point compared to 400 per watering point today. It was during this period that large changes in vegetation composition were thought to have occurred (Beadle, 1948; Perry, 1977).

The country was grazed also by rabbits. They invaded the west Darling in the early 1880s and spread to the Queensland border within ten years. In 1887 rabbits were being destroyed on Kinchega at the rate of 100,000 per month for several months with 'no noticeable difference' (Western Lands Commission records). In a letter to the Western Lands Commission in 1906, the manager of Netley (now Tandou) wrote: 'Five years ago I travelled through Netley and for half the journey (40 miles) there was not a blade of grass along the whole stage, nor could a living bird or beast or living shrub or bush be seen, a desolate dead land, eaten out, everything that could grow killed by the rabbits. This year there is a great change, grass fairly plentiful, shrubs sprouting and flowers.' Abundant rabbits competed with sheep for vegetation until the introduction of myxomatosis in the 1950s (Fenner & Ratcliffe, 1965). Since then they have declined in both numbers and distribution (Fenner & Myers, 1978).

Heavy grazing by sheep and rabbits was compounded by drought. Severe droughts have occurred at Menindee in every decade since the 1860s with the exception of the 1950s and 1970s. The decade of 1895-1904 is the driest recorded (Table 2.3). It coincided with the highest numbers of sheep on record and a high density of rabbits. In that decade sheep numbers in western New South Wales dropped by over 50% (Butlin, 1962).

This period of heavy grazing and drought in the late 1800s and early 1900s coincided with the disappearance from western New South Wales of several small and medium-sized marsupials. However, large species of kangaroos increased with the advent of sheep. There were few kangaroos when the region was settled but by the late 1800s they were deemed pests (E. Rolls, pers. comm.). Krefft (1871) considered the 'prolific increase' in large kangaroos was explained by 'the destruction of the Native Dog and the absence of the aboriginal hunting.' While dingo control and the demise of aboriginal hunting practices undoubtedly played a part, the establishment of watering points for stock was probably more influential. This practice continued through to the 1920s and 1930s when properties were subdivided into paddocks each with permanent water.

What were the effects of this history of grazing on the vegetation of the sheep rangelands? At settlement the Riverina of New South Wales was open woodland of myall (*Acacia pendula*) with a well developed understorey of old man saltbush (*Atriplex nummularia*) (Moore, 1953a). Their disappearance is attributed to the severe drought of 1875-7 and to grazing by sheep. In drought myall was felled for stock feed. It has not regenerated under grazing. Old man saltbush now remains only as remnants in areas that were not grazed. These species were replaced by the present grassland of wallaby grass (*Danthonia* spp.) and speargrass (*Stipa falcata*), possibly through an intermediate stage of bladder saltbush (*Atriplex vesicaria*).

Bladder saltbush occurs over much of arid and semi-arid Australia. It is a long-lived perennial shrub, highly tolerant of drought, but it is intolerant of heavy grazing (Wilson, Leigh & Mulham, 1969). Forty-five years of grazing by stock and rabbits severely reduced its density in a large area of South Australia (Fatchen, 1978). It was replaced in the Riverina by *Danthonia caespitosa* grassland and by copper burr (*Sclerolaena* spp.) and annual saltbush (*Atriplex* spp.) in other districts (Williams, 1974).

Kangaroo grass (*Themeda australis*) was one of the most widely distributed species of the Australian flora, extending from the arid inland to the alps (Cunningham *et al.*, 1981). It has largely disappeared under grazing by sheep and is now found only in ungrazed areas such as cemeteries, railway reserves and on little-used stock routes. Heavy grazing of this summer-growing perennial grass has led to its replacement by winter-growing native and exotic species (Moore, 1953b).

Since settlement exotic grasses and forbs have invaded all plant communities in the sheep rangelands. Approximately 20% of the plants in Western New South Wales are exotic (Cunningham *et al.*, 1981). They were introduced either by accident or as fodder plants. Most are annual forbs and grasses in the families Fabaceae and Poaceae. They grow in winter and spring unlike some of the summer-growing native species that they have either confined or replaced (Moore, 1959; Williams, 1961).

Woody shrubs increased in the semi-arid woodlands of New South Wales after settlement and reduced carrying capacity for stock (Harrington *et al.*, 1979). Woody shrubs are favoured by current management practices of heavy grazing and suppression of wild fires. *Acacia* spp., *Eremophila* spp. and *Dodonaea* spp. increase under grazing but are either killed or reduced by fire (Harrington *et al.*, 1984). Grazing prevents growth of grasses to a biomass providing enough fuel to carry wildfire over large areas.

Recruitment of some long-lived species of trees and shrubs has been inadequate for replacement this century despite favourable climatic conditions. For example, seedlings of western myall (*Acacia sowdenii*) in the rangelands of South Australia establish only a few times each century (Lange & Purdie, 1976). They are killed when grazed by sheep and rabbits. Similarly, recruitment of *Acacia burkittii* is supressed by sheep and rabbits (Crisp & Lange, 1976). Both species seem fated to disappear under current stocking practices.

2.42   *The sheep rangelands*

The vegetation communities in the sheep rangelands consist of open forest, woodlands, shrublands and grasslands (Fig. 2.5). There is a wide overlap between communities in the composition of pastures, so only the tallest stratum is considered here. Descriptions are from Moore & Perry (1970), Moore (1975) and Specht (1973). The woodland and shrubland communities can be subdivided further into several distinct types.

Open forest is restricted to the banks of the major rivers and their billabongs. It is dominated by the river red gum (*Eucalyptus camaldulensis*).

Woodlands include associations of poplar box (*Eucalyptus populnea*) and cypress pine (*Callitris* spp.) and areas on floodplains that are dominated by black box (*E. largiflorens*), coolibah (*E. microtheca*) or yapunyah (*E.*

Fig. 2.5. The major vegetation associations of the sheep rangelands.

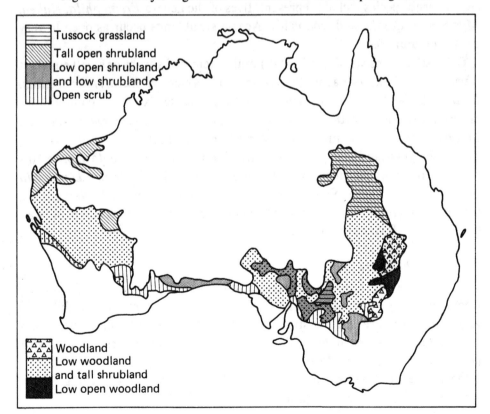

Tussock grassland
Tall open shrubland
Low open shrubland and low shrubland
Open scrub

Woodland
Low woodland and tall shrubland
Low open woodland

ochrophloia). Poplar box and cypress pine may occur either as pure stands or together. The shrub understorey includes lignum (*Muehlenbeckia cunninghamii*) and *Chenopodium* spp. in the former community, and species of *Eremophila, Cassia* and *Dodonaea* in the latter.

Low woodlands and low-open woodlands are dominated by either the belah (*Casuarina cristata*) — rosewood (*Heterodendrum oleifolium*) association or cypress pine. The former overtops shrublands of *Eremophila* spp., *Dodonaea* spp., *Cassia* spp. and, in places, *Maireana* spp. and *Acacia* spp.

Low shrubland and low-open shrubland occupy about 10% of the sheep rangelands and are dominated by members of the family Chenopodiaceae (saltbushes and bluebushes), such as black bluebush (*Maireana pyramidata*), pearl bluebush (*M. sedifolia*) and various species of *Atriplex*. All are xerophytic halophytes. These shrubs form extensive stands and may be up to about 1.5 m in height.

Tall and tall-open shrublands occupy about 50% (850,000 km^2) of the sheep rangelands. They are 2-8 m in height and are dominated by mulga (*A. aneura*). Gidgee (*A. cambagei*), western myall (*A. sowdenii*), yarran (*A. homalophylla*), and witchetty bush (*A. kempeana*) are also prominent. Understorey shrub species include representatives of the genera *Eremophila, Hakea, Myoporum, Cassia*, and *Dodonaea*. Acacia shrublands occur as open to very dense communities.

Open scrub consists of low ($\langle$12 m) multi-stemmed shrubs or small eucalypt trees commonly known as mallee. Dominant species are *Eucalyptus socialis, E. dumosa* and *E. oleosa*. They form very dense extensive communities often in association with understorey shrubs (*Cassia* spp., *Eremophila* spp., *Myoporum* spp.) or hummock grassland (*Triodia* spp.).

Tussock grasslands are principally the mitchell grass (*Astrebla* spp.) plains of the Northern Territory and Queensland, and the disclimax *Danthonia* grassland of southern New South Wales. *Astrebla* grasslands are usually treeless. The ground between tussocks may be bare in dry times and populated by annuals in favourable seasons.

2.43   *The study area*

The vegetation on Kinchega and Tandou consists of open forest, woodland and shrubland (Fig. 2.6). There is a high degree of variability in vegetation structure and floristics and this reflects wide differences in species composition, soils and micro-relief and the interaction of these factors with herbivores, weather and flood.

Open forest is confined to the banks and billabongs of the Darling River. It is dominated by river red gum.

Woodlands, low woodlands and low-open woodlands occur across the floodplain, depressions, drainage tracts and on the lake margins. They are dominated by black box. These woodlands vary in density and height and contain a mosaic of vegetation communities in their substoreys. Those regu-

Fig. 2.6. The vegetation of Kinchega National Park and Tandou Property.

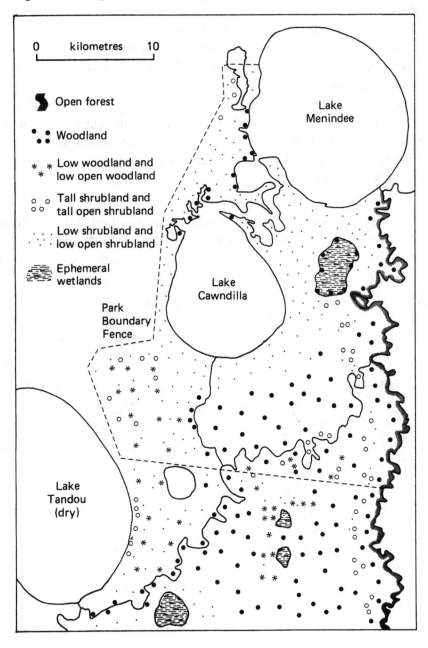

larly flooded have a dense understorey of nitre goosefoot (*Chenopodium nitrariaceum*), lignum and canegrass (*Eragrostis australasica*) with a ground stratum of black roly-poly (*Sclerolaena muricata*). Those flooded less often may have large open flats of chenopod sub-shrubs, grasses and forbs. Low woodland and low-open woodland also occur on sandplains and dune crests where associations of rosewood — belah, *Acacia* spp. and *Hakea* spp. occur as dominants.

Tall shrublands and tall-open shrublands are scattered over parts of the sandplain and higher rises of the floodplain. They comprise clumps of prickly wattle (*Acacia victoriae*), needlewood (*Hakea leucoptera*) and turpentine (*Eremophila sturtii*).

Low shrublands and low-open shrublands occur as dense stands or as scattered bushes across most of the sandplains and dunefields. The dominant shrub is black bluebush with turpentine, hopbush (*Dodonaea attenuata*), *Cassia* spp., and spotted fuschia (*Eremophila maculata*) occurring as co-dominants in some areas.

Fig. 2.7. View from the air of the black box woodland of the floodplains of western New South Wales. Here it is partially flooded.

Pastures, in all the communities discussed above, fluctuate markedly in biomass and species composition in response to a complex array of biotic and abiotic factors such as rainfall, soil texture and structure, nutrients, microtopography, soil temperature, interspecific plant competition, and intensity and type of herbivory. Winter rain promotes the growth of annual grasses, annual forbs, and chenopod sub-shrubs whereas summer rain favours these plus perennial grasses. In general a different suite of plants grows in summer than in winter, although there is a wide seasonal overlap with some species. Of 290 pasture species, 69% are forbs, 15% are grasses, 14% are woody sub-shrubs and 2% are sedges, rushes or ferns. Forty-two percent of pasture species are annuals, 44% are perennials, and 14% have annual, perennial or biennial reproductive strategies depending on seasonal conditions. Annual and perennial plants made up approximately 60% and 40% respectively of estimated biomass of the pasture during the study. Approximately 20% of the grasses and 20% of the forbs are exotics.

Most annual species evade dry periods and drought by surviving as seed. They grow only when conditions are favourable. Perennial pasture species, principally sub-shrubs such as *Atriplex spp.* and *Sclerolaena spp.*, persist through all but the driest seasons.

Fig. 2.8. The black box woodland with its understorey of chenopod shrubs.

2.5   Fauna

The fauna of the sheep rangelands includes about 30 species of amphibians, 220 species of reptiles, 200 species of birds and 80 species of mammals. Many of the birds are common and conspicuous, particularly the parrots, the emu, the ground-feeding pigeons and doves, and a large number of raptors. Most of the mammals are small and nocturnal and are seldom seen. The large kangaroos are more conspicuous, being common throughout the rangelands after a run of good years and less common after drought.

Reptiles are the richest in species. Seventy-five percent of these are lizards, 23% are snakes and the remainder are turtles. The high number of lizard species in arid and semi-arid Australia has been attributed variously to a usurping of the ecological roles of snakes, insects, worms and mammals (Pianka, 1969), the high structural diversity of habitat created by hummock grasses (*Triodia* spp. and *Plectrachne* spp.), 'temporal partitioning' of the environment (i.e. the presence of many nocturnal forms of geckoes and burrowing skinks), and extensive radiations of a few genera (e.g. *Amphibolurus* (dragons) and *Ctenotus* and *Lerista* (skinks)) (Cogger, 1984).

Fig. 2.9. An adult male red kangaroo in front of prickly wattle shrubs and black box woodland.

Particular locations within the rangelands hold less than the full fauna. Kinchega has about 190 bird species, 50 species of reptiles and amphibians and 27 species of mammals. The bird fauna contains a large complement of water birds owing to the habitat provided by permanent lakes and the Darling River.

Diverse as the present fauna is, it was much more diverse at the time people discovered Australia. At that time, more than 30,000 years ago, the fauna of the rangelands contained many more species and many more large forms. A prehistoric record of the composition of the mammalian fauna of the Menindee area is provided by bones in the fossil sand dunes on the eastern margins of Tandou, Cawndilla and Menindee Lakes. Mammal species present then, but absent at the time of European settlement, include a rhinoceros - like marsupial (*Diprotodon*), numerous species of large kangaroos (*Sthenurus* spp., *Procoptodon*, *Protemnodon*), the large wombat (*Phascolonus*), the koala (*Phascolarctos*), the small wallabies and rat kangaroos (*Propleopus, Caloprymnus, Lagorchestes conspicillatus*), and the predators (*Thylacoleo, Thylacinus* and *Sarcophilus*) (Tedford, 1967; Marshall, 1973; Merrilees, 1973; Hope, 1978).

Fig. 2.10. A female (left) and male (right) western grey kangaroo in woodland.

Europeans and their sheep and cattle moved into the rangelands of western New South Wales during the 1830s. The expedition of Blandowski in 1856-7 collected information on the fauna of an area of about 300 km radius around the junction of the Murray and Darling Rivers. Wakefield (1966) summarised the information recorded by Gerard Krefft, a member of the expedition who left the only usable records. The fauna at that time included a monotreme, 22 marsupials and seven rodents. The monotreme is still common, but 16 of the marsupials and five of the rodents have long since disappeared from the region. Some, such as the pig-footed bandicoot (*Chaeropus ecaudatus*) and the eastern hare-wallaby (*Lagorchestes leporides*), are extinct. Some such as the burrowing bettong (*Bettongia lesueur*) and the greater stick-nest rat (*Leporillus conditor*) are now confined to small offshore islands. Others, such as the bridled nail-tailed wallaby (*Onychogalea fraenata*) and the bilby (*Macrotis lagotis*), are extremely rare. They are confined to small areas of the mainland that constitute very small fractions of their former ranges. It seems likely that the alteration of the habitat by stock, competition for food and warrens by the introduced rabbit, and predation by the introduced fox

Fig. 2.11. The bilby, a small burrowing bandicoot of about 2 kg, occurred throughout arid and semi-arid Australia at the time of European settlement. It is confined now to isolated localities in Central Australia.

and cat have all contributed to the decline of this suite of mostly small, ground-dwelling mammals.

The mammalian fauna of the sheep rangelands at the time of European settlement (Table 2.4) was divided equally between marsupials (38 species) and eutherians (45 species of bats and rodents). Today 32% of the former and 13% of the latter are extinct within the rangelands. The extinct marsupials include all of the bandicoots and six of the 14 macropods. All of the eutherian extinctions were rodents and included both species of stick-nest rats (*Leporillus*) and a number of species of *Pseudomys* (e.g. *P. gouldii*, *P. desertor*). Eutherians introduced since European settlement are abundant, widespread, and generally regarded as pests. They include the rabbit, fox, cat, feral dog, feral pig and feral goat, house mouse and black rat. Because of their widespread distribution, in contrast to that of many endemic species, they dominate the fauna more than is suggested by Table 2.4. For example, mammals introduced by Europeans represent 30% of the mammalian fauna at Kinchega.

Fig. 2.12. The burrowing bettong, a small rabbit-sized wallaby, occurred throughout arid and semi-arid Australia. It is now found only on four islands off the coast of Western Australia. Competition with rabbits for food and warrens, predation by introduced cats and foxes, and a change in vegetation biomass and structure caused by stock grazing, are probably the cause of its decline.

The net change in the fauna of the sheep rangelands in the last 150 years is a reduced diversity of marsupials (particularly of small species dependent on low vegetation for cover) and an increase in the diversity of eutherians. Changes wrought by pastoralism have inadvertently benefited some marsupial species. The extirpation of the dingo and the provision of watering points for sheep have increased the numbers of the larger macropods.

Table 2.4. *The relative status of mammalian groups occurring in the sheep rangelands. Only species whose ranges extend well into the sheep rangelands are included. Mammals which occurred within the area at the time of European settlement but not now present are included also. Data are from Strahan (1983). Status refers to that within the sheep rangelands.*

Mammal group	Common/ abundant	Uncommon/ rare	Extinct	Total	
Monotremes	1	—	—	1	
Sub-total					1
Marsupials (Super family)					
Carnivores/Insectivores (Dasyuroidea)	6	6	3	15	
Bandicoots (Perameloidea)	—	—	3	3	
Koala/wombats (Vombatoidea)	1	1	—	2	
Possums/gliders (Phalangeroidea)	4	—	—	4	
Kangaroos/wallabies (Macropodidea)	7	1	6	14	
Sub-total					38
Eutherians (Order)					
Bats (Chiroptera)	13	6	—	19	
Endemic rats/mice (Rodentia)	6	4	5	15	
Introduced	10	—	1	11	
Sub-total					45
Total					84

3

The effect of weather on soil moisture and plant growth in the arid zone

GREG WELLARD

3.1   Introduction

Rainfall is the dominant factor controlling the growth of plants in arid lands throughout the world (Beatley, 1969; Ross, 1969, 1976; Noy-Meir, 1973; Ross & Lendon, 1973; Gutterman, 1981; Orr, 1981). Much of the variability of pasture biomass can be explained by rainfall (Noy-Meir, 1973; Chapter 4). Temperature also plays a role in the initiation and success of germination and the survival of seedlings (Beatley, 1967; Gates & Muirhead, 1967; Lendon & Ross, 1978). Hence, rain falling at different seasons in the arid zone stimulates the germination of different suites of plant species.

This chapter reports the effect of rainfall and temperature on changes in soil moisture, pasture biomass, pasture growth and species composition on the two major soil types of Kinchega National Park.

The study posed the following questions:

(1) What is the relationship between rainfall and soil moisture and how does it differ between the two major soil types within the Park?

(2) Can trends in soil moisture or a knowledge of rainfall and temperature be used to predict changes in pasture growth and biomass?

The study was conducted within two exclosures, one on each of the two major soil types within Kinchega National Park. A fence excluded mammalian herbivores. Hence, growth, dieback and trends in biomass are measured in the absence of grazing. There is considerable evidence that grazing affects the growth rate of plants, sometimes stimulating growth and sometimes depressing growth or killing the plant (Harper, 1977; McNaughton, 1979). This study differs from that in Chapter 4 in that vegetation parameters are measured at greater frequency (monthly v. 3-monthly) but are restricted to two sites only compared to the 313 sites monitored across Kinchega and the adjoining sheep station.

3.2 Study exclosures

Two areas of 250 m² were chosen for study. Each was broadly representative of one of the two major soil types within the Park: the light-textured aeolian sands which characterise the dunefields and sandplains, and the heavy-textured alluvial loams which characterise the floodplain (Fig. 2.4).

3.21 *Soil*

The floodplain exclosure had a soil of alluvial clay-sandy-loam to a depth of approximately 2 m. The sandplain exclosure had an aeolian sandy-clay-loam to a depth of approximately 2 m. Table 3.1 gives details of both soil profiles.

The ratio of sand to clay was about 1:1 for the floodplain soil and 4-6:1 for the sandplain soil. The floodplain soil, with a clay content 2.5 times that of the sandplain soil, had a higher water-holding (field) capacity, a higher wilting point and a lower bulk density (Table 3.2). Although the floodplain soil holds more moisture, this water is tightly bonded to the clay particles and less accessible to plants. Both sites were flat so that ponding and run-off were slight. Deep soil drainage was enhanced at both sites by buried aeolian sand below 2 m. Both soils contained carbonate nodules within diffuse layers or less commonly as a single layer between 20 and 50 cm below the surface.

3.22 *Vegetation*

The vegetation at the floodplain site was a mixture of perennial sub-shrub species less than 0.5 m high of the genus *Sclerolaena* (Chenopodiaceae). Between these perennial sub-shrubs were a variety of annual sub-shrubs, forbs and grasses, including species of *Atriplex* (Chenopodiaceae), *Craspedia* and *Helipterum* (Asteraceae), *Tetragonia* (Aizoaceae) and *Dactyloctenium*

Table 3.1. *Structural analysis of the soils in the study exclosures.*

Site	Depth (cm)	Structural Analysis (%) Clay	Silt	Sand	Gravel	pH
Flood-plain	0–50	48	11	41	0	8.3
	50–100	27	37	36	0	9.0
	100–200	51	8	41	0	9.0
	200+	Buried aeolian sand – composition unknown				
Sand-plain	0–5	13	3	82	2	8.5
	5–200	21	3	74	2	8.5
	200+	Buried aeolian sand – composition unknown				

(Poaceae). The composition of this pasture layer changed with the intensity and seasonality of previous rainfall.

The sandplain site supported a vegetation of medium to large perennial shrubs (1-3 m high) of prickly wattle (*Acacia victoriae*) and bluebush (*Maireana pyramidata*), with blackbox (*Eucalyptus largiflorens*) trees at low densities. The pasture consisted of perennial sub-shrubs less than 0.5 m high of *Brachycome* and *Ixiolaena* (Asteraceae), *Sclerolaena*, *Bassia*, *Chenopodium*, *Atriplex* (Chenopodiaceae) and forbs and grasses including species of the genera *Craspedia* and *Helipterum* (Asteraceae), *Zygophyllum* (Zygophylla-ceae), *Plantago* (Plantaginaceae) and *Enneapogon* (Poaceae). Again the composition of the pasture at any time depended on the amount and seasonality of previous rainfall. The sandplain site had more species than the floodplain site (Table 3.3).

Table 3.2. *Water-holding capacity (% moisture) and bulk density of soils within the study exclosures.*

Soil type	Bulk density	Wilting point	Field capacity
Floodplain	1.4	16	30
Sandplain	1.6	5	12

Analysis: New South Wales Soil Conservation Service.

Table 3.3. *Species richness of the pasture in the two study exclosures.*

| Site | Number of species represented | | |
	Perennial sub-shrubs	Forbs	Grasses
Floodplain	5	20	7
Sandplain	9	28	8

3.3   Methods

Measurements of weather parameters and pasture biomass were made from February 1981 to February 1984.

3.31   *Weather measurements*

A meteorological station, established at a central point 2 and 5 km respectively from the floodplain and sandplain sites, recorded:

temperature   — weekly maximum and minimum and a continuous record of temperature from a thermograph.

rainfall   — weekly with a bulk-storage raingauge and a continuous record from a syphoning pluviograph.

evaporation   — weekly from a Class A type evaporation pan.

Rainfall was also measured directly at the two exclosures.

3.32   *Soil moisture*

Soil moisture was measured weekly at 10 cm intervals of depth to 50 cm. Determinations below 10 cm were pooled because values were not significantly different.

3.33   *Pasture biomass*

Pasture biomass was estimated each month by the dry weight rank and comparative yield techniques of Mannetje & Haydock (1963), Haydock & Shaw (1975) and Jones & Hargreaves (1979). Biomass on each soil type was estimated from 30 quadrats each 0.5 m². The biomass of each pasture species was assessed separately and was bulked later into broad categories of growth habit.

Instantaneous growth rate was calculated as the difference between the natural logarithm of successive four-weekly biomasses. A doubling of plant biomass from 10 to 20 kg/ha gives the same rate as a doubling from 500 to 1,000 kg/ha.

3.4   Results

3.41   *Weather*

Rainfall was below average in the summer of 1980-81, the spring of 1981 and the summer of 1981-82. Dry conditions changed to drought with the failure of the rains in the winter and spring of 1982 and the summer of 1982-83. Rain during this period was insufficient for germination and growth. Exceptionally good seasons followed the breaking of the drought in March 1983.

The high spatial variability of rainfall characteristic of arid lands (e.g. Sharon 1972a, b, 1979) was also found at Kinchega. Figure 3.1 indicates that, although the two study sites were only 4 km apart, in 40% of the four-weekly intervals rainfall differed by at least 25%.

Mean maximum and minimum temperatures during the study period are presented in Fig. 3.2 for each four-weekly interval. Mean maximum daily temperatures exceeded 40°C in summer of each year and mean minimum temperatures fell to around 0°C in winter.

Fig. 3.1. Rainfall at each study exclosure in 4-weekly intervals.

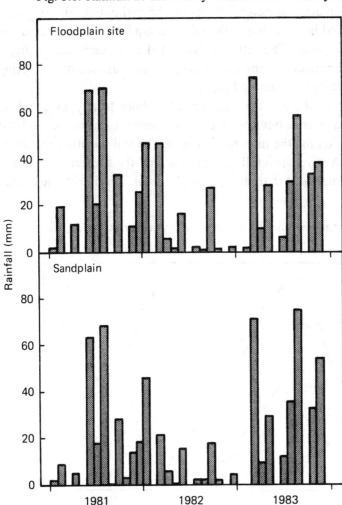

3.42   Soil moisture

Trends in soil moisture for the surface and the sub-surface strata for each soil type are shown in Fig. 3.3. They are similar. However the moisture content of the floodplain soil in all months, with the exception of June 1983, was considerably higher than that of the sandplain soil. This consistent difference is a reflection of the composition of the soils, the floodplain soil having a higher clay content and therefore a greater capacity to retain moisture. Concomitant with the higher capacity of clay soils to retain water in their profile is their correspondingly higher wilting point (Table 3.2). Hence, although the floodplain soil contained more moisture than the sandplain soil, moisture content was below the estimated wilting point for a longer time. The 10-50 cm layer was below wilting point in late autumn and early winter in 1982 and again during the height of the drought of 1982 and early 1983. The moisture in the sandplain soil never fell below the estimated wilting point. Some caution is needed in interpreting the relationship between measured soil moisture and wilting point. The latter is estimated in a laboratory using a standard species of temperate plant which may be less capable of extracting water from the soil than arid-adapted plants.

Soil moisture (θ_v) in the surface and subsoil of both soil types could be predicted by regression equations with two independent variables — total rainfall and evaporation for the month previous to the soil moisture measurements (Table 3.4). A given rainfall produces a slightly greater rise in soil moisture in the floodplain soil than in the sandplain soil. Similarly, the deeper

Fig. 3.2. Mean maximum and minimum daily temperatures at Kinchega National Park.

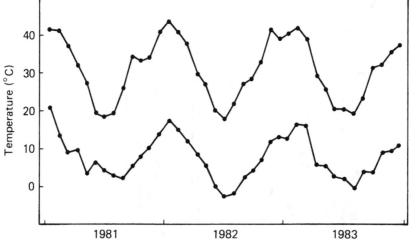

floodplain soil tended to lose moisture by evaporation at a slightly greater rate than did the deeper sandplain soil.

3.43   *Pasture growth rate*

The instantaneous growth rates of the chenopod sub-shrubs, forbs and grasses were highly dynamic, frequently switching from positive (growth) to negative (dieback) (Fig. 3.4). Growth rate was much more labile than pasture biomass (Fig. 3.5), which reflected the history of growth rates over several months.

Perennial sub-shrubs on both soil types and grasses on the sandplain showed the lowest variance in growth. They rarely exceeded a growth rate of ± 0.649 (i.e. a doubling or halving of biomass in a four-weekly period) and appeared to be buffered against dieback resulting from high summer temperatures. In contrast forbs grew fast and died back rapidly, particularly on the sandplain, as did the grasses of the floodplain. Forbs on the sandplain grew at rates of 1.5-2.5 (equivalent to an increase of between 500 and 1,000%) in a four-weekly period in the autumn of 1982 and 1983 and declined at a similar rate in the summers of 1981–82 and 1982–83.

Fig. 3.3. Trends in soil mosture at 0-5 cm and 10-50 cm at the floodplain and sandplain sites.

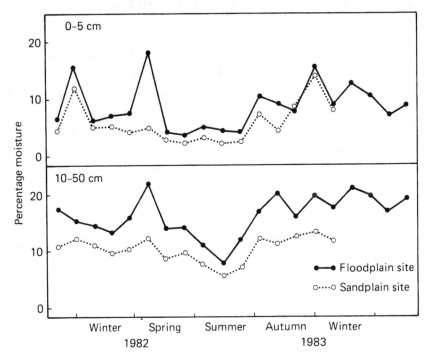

The period of maximum plant growth generally occurred 2-3 months earlier than the peak in biomass. For example, forbs on the sandplain peaked in biomass in spring but their highest instantaneous growth was in autumn (1982, 1983) and winter (1981). Similarly, grasses on the floodplain attained peak biomasses in the autumn of 1982 and 1983 but their major growth periods were in late summer.

Growth of all plant groups was correlated best with rainfall of either the previous one or two months (Table 3.5) in contrast to biomass which correlated most highly with rainfall over much longer time periods (six months for forbs, 12 months for perennial sub-shrubs, and 12 months for grasses on the sandplain). Growth of a particular plant group was weakly (and usually not significantly) negatively correlated with biomass of that plant group implying a weak parabolic relationship between growth increment and plant biomass, growth increment being highest at intermediate biomasses.

Growth of some plant groups showed a strong relationship with temperature. Forbs on both soil types, and perennial sub-shrubs on the sandplain showed a significant negative correlation with maximum temperature suggesting a predominantly winter growth period. Growth of the summer-growing annual grass *Dactyloctenium radulans* on the floodplain soil showed the reverse pattern. Growth of grasses on the sandplain and perennial sub-shrubs on the floodplain showed no relationship with temperature. The pattern of growth for each plant group by soil type is examined in detail below.

Sandplain soil type
Forbs grew in the autumn and winter of all three years of the study. Initiation of growth appeared to result from a high monthly rainfall (65 mm in June 1981, 45 mm in January 1982, and 75 mm in March 1983). Growth flushes

Table 3.4. *Soil moisture in mm (θ_v) as a function of rainfall over the previous month (x_1) and evaporation over the previous month (x_2).*

Soil type	Soil depth	
Floodplain	0-5 cm	$\theta_v = 0.074\, x_1 - 0.192\, x_2 + 10.606$ $(r^2 = 0.435)$
	10–50cm	$\theta_v = 0.071\, x_1 - 0.184\, x_2 + 17.935$ $(r^2 = 0.511)$
Sandplain	0-5cm	$\theta_v = 0.058\, x_1 - 0.190\, x_2 + 8.191$ $(r^2 = 0.535)$
	10-50cm	$\theta_v = 0.058\, x_1 - 0.141\, x_2 + 12.05$ $(r^2 = 0.784)$

Fig. 3.4. Instantaneous growth rates of perennial sub-shrubs, forbs and grasses on floodplain and sandplain soils.

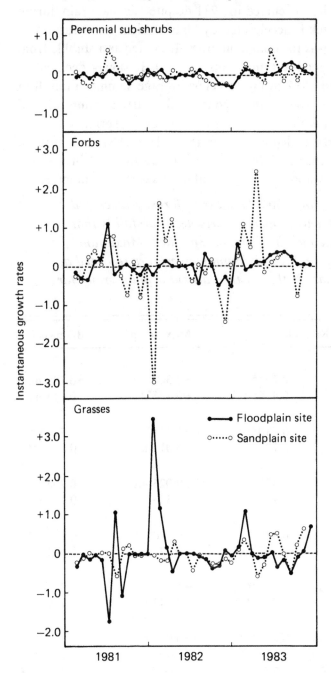

were followed by post-flowering senescence in spring and summer. Senescence and dieback of forbs may be independent of rainfall during periods of extreme ambient temperatures such as occurred in 1981 despite adequate rain during spring and summer. Dieback is accelerated by high ambient temperatures.

Enneapogon avenaceus was the dominant grass. It existed as a stubble from a growth flush following good autumn rains in 1980. It slowly declined in biomass through 1981 and 1982 owing to below average summer rainfall in these years. It commenced growing in response to the early autumn rains of 1983 but its major growth did not occur until the following spring.

Perennial sub-shrubs responded to winter rainfall in 1981 and autumn rainfall in 1983. In both cases rainfall in excess of 60 mm fell in a single month and was followed by above average rainfall in succeeding months.

Table 3.5. *Correlation of growth and biomass of forbs, grasses and perennial sub-shrubs with environmental variables. Rainfall is that recorded at the site for the month or months specified. Maximum temperature is the mean monthly maximum temperature recorded at the weather station. Biomass is that of the particular plant group for the previous month.*

Soil type		Rainfall	Max. temp.	Biomass
Forbs – Growth				
Sandplain	prev. mth	0.298*	−0.300*	−0.182 n.s.
Floodplain	prev. mth	0.456**	−0.395**	−0.115 n.s.
Forbs – Biomass				
Sandplain	prev. 6mths	0.630***	−0.348*	0.773***
Floodplain	prev. 6mths	0.722***	−0.013 n.s.	0.812***
Grasses – Growth				
Sandplain	prev. mth	0.375*	−0.025 n.s.	−0.295*
Floodplain	prev. mth	0.209 n.s.	0.326*	−0.161 n.s.
Grasses – Biomass				
Sandplain	prev. 12mths	0.425**	−0.170 n.s.	0.901***
Floodplain	prev. 6mths	−0.102 n.s.	−0.226 n.s.	0.841***
Perennial Sub-shrubs – Growth				
Sandplain	prev. 2mths	0.387**	−0.327*	−0.258 n.s.
Floodplain	prev. mth	0.466**	−0.247 n.s.	−0.179 n.s.
Perenial Sub-shrubs – Biomass				
Sandplain	prev. 12mths	0.603***	−0.038 n.s.	0.820***
Floodplain	prev. 12mths	0.539***	−0.076 n.s.	0.921***

* significant at $P\langle0.05$, ** significant at $P\langle0.01$, *** significant at $P\langle0.001$, n.s. not significant.

Floodplain soil type

Forbs grew in response to winter rains in 1981 and autumn and winter rains in 1983. Both their growth and dieback were at a slower rate than forbs on the sandplain. Initiation of growth appeared to result from a high monthly rainfall as on the sandplain soil (65 mm in June 1981, and 75 mm in March 1983). In contrast to forbs on the sandplain soil they did not show a marked growth response to the 45 mm of rain falling in March 1982.

Grasses (mainly *Dactyloctenium radulans*) grew strongly following the summer rainfall of 1981-82 (Dec. 25 mm, Jan. 45 mm) setting seed in early March (Autumn) 1982. This growth was not sustained despite rain in March (45 mm). This species showed similar growth spikes in March 1983 in response to a large fall of rain in that month (75 mm) and in December 1984 in response to above average spring rainfall.

The biomass of the chenopod sub-shrubs was either static or declining despite good rainfall in winter 1981 and the summer of 1981-82. Sub-shrubs grew slowly following autumn rains in 1983.

3.44   *Pasture biomass*

The biomass of perennial sub-shrubs, grasses and forbs in each enclosure over the study period is shown in Fig. 3.5. The perennial sub-shrubs showed the greatest absolute change in biomass, declining from 1,100 to 300 kg/ha on the floodplain and from 400 to 130 kg/ha at the sandplain site. Forbs on the sandplain showed abrupt peaks in biomass in the spring of 1981 and the winter and spring of 1983. No winter-spring flush occurred in 1982 because of insufficient rain. Forbs on the floodplain showed a similar but more subdued pattern with a small rise in biomass in early spring of 1981, no flush in 1982, and a flush in the spring of 1983, 3-4 months later than that for forbs on the sandplain. Grasses on the sandplain (mainly *Enneapogon avenaceus*) showed a slow attrition of biomass from February 1981 through to the breaking of the drought in March 1983. They reacted weakly to the drought-breaking rains of autumn 1983, not making major increases in biomass until late spring and early summer of 1983.

The annual grass, *Dactyloctenium radulans*, on the floodplain soil increased in biomass in the late summer of 1981-82 and autumn 1983 in response to summer and early autumn rainfall.

The total vegetation biomass on both soil types regressed significantly on rainfall over the previous 12 months (Fig. 3.6). The ungrazed biomass predicted for the annual average rainfall at Kinchega was 850 ± 65 kg/ha (estimate $\pm$ 95% confidence interval) on the floodplain soil and 713 ± 52 kg/ha on the sandplain soil.

3.45   Species composition

Annual forbs comprised approximately 60% of the species on both soil types. On the floodplain soil the majority of annual forbs (60%) germinated in the cooler months. Thirty percent germinated in the intermediate months, and the remainder (10%) in the warmer months. On the sandplain soil 66%, 9%, and 25% of annual forbs germinated in the cooler, intermediate and warmer months respectively. Grass species exhibited a different response to

Fig. 3.5. Trends of biomass of perennial sub-shrubs, forbs and grasses on floodplain and sandplain soils.

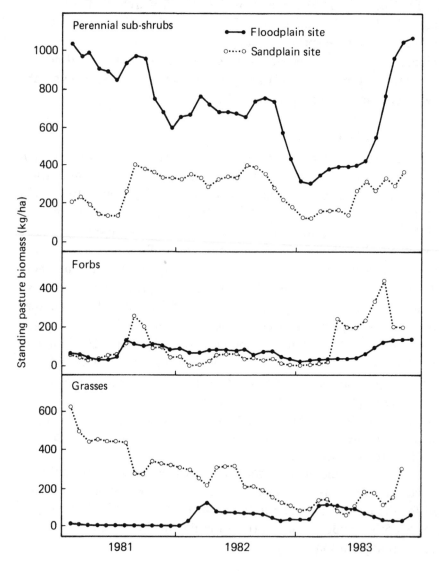

season on each soil type with the majority of species on the floodplain soil germinating in the warmer months (44%) and a majority on the sandplain soil germinating in the cooler months (55%). Hence the community on the sandplain contained a higher proportion of warm season forbs and a lower proportion of warm season grasses.

Fig. 3.6. Trend of pasture biomass (V) against rainfall for the previous 12 months (R_{12}). For floodplain soil $V = 246.2 + 2.56R_{12}$, $r^2 = 0.46$, $P<0.001$ and for sandplain soil $V = 264.5 + 1.90R_{12}$, $r^2 = 0.38$, $P<0.001$.

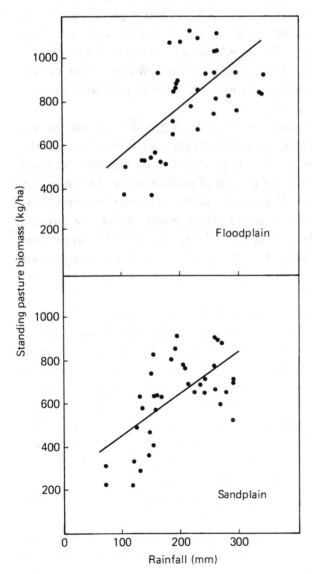

3.5   Discussion

Instantaneous growth of forbs, grass and perennial sub-shrubs were most highly correlated with rainfall in either the previous month or previous two months. Immediate response to rainfall is a feature of arid-zone plants, particularly annuals. Mott & McComb (1975) reported that annual grasses and composites in the Murchison area of Western Australia in a 200 mm rainfall zone took 24-27 days from a rainfall event to the initiation of flowering. The particular suite of plants which germinated depended on the seasonality of rainfall. Winter rainfall stimulated the germination of forbs (*Helichrysum cassinianum* and *Helipterum craspedioides*) and summer rainfall initiated the germination of grasses (*Aristida contorta*). A similar pattern is seen at Kinchega. Rainfall in autumn and winter initiated the growth of forbs (e.g. *Craspedia* spp., *Helipterum* spp., *Tetragonia tetragonioides* on the sandplain, and *Helipterum floribundum*, *Lotus cruentis*, *Carrichtera annua* on the floodplain). The annual grass *Dactyloctenium radulans* responded to summer rainfall.

Growth at Kinchega was initiated by rainfalls of greater than 40 mm per month in any season. In a chenopod shrub community in South Australia Noble (1977) found that summer falls of greater than 85 mm were required to initiate germination and a pulse of growth of pasture species (*Stipa nitida*, *Erodium cygnorum*, *Salsola kali*, *Bassia* spp. and *Zygophyllum* spp.). He concluded that irregular but frequent wet summers determined the biomass dynamics of the vegetation despite that area being in a winter rainfall zone. In the Mojave and Great Basin Deserts of southern Nevada large-scale germinations resulted from only 25 mm of winter rainfall (Beatley, 1967, 1969).

Total pasture biomass on both soil types in the study exclosures at Kinchega was predicted best by rainfall over the previous 12 months. Each 1 mm of annual rain produced an additional 2 kg/ha of biomass. Similarly, the biomass of pasture species in a chenopod shrubland (*Atriplex vesicaria* and *Maireana sedifolia*) was most highly correlated with rainfall over the previous 12 months in South Australia (Noble, 1977). The pasture that he examined was ungrazed and contained a mix of perennial sub-shrubs and annual grasses and forbs similar to that at Kinchega. The biomass of pasture ungrazed for the duration of the project (this study) fell to 225 kg/ha (sandplain) and 375 kg/ha (floodplain) in the final stages of the drought. Equivalent areas outside the enclosures had a pasture biomass of 0 kg/ha (sandplain) and 40 kg/ha (floodplain).

Davidson & Donald (1958) reported upwardly convex and parabolic relationships for plant growth increment as a function of biomass for sown pastures in higher rainfall areas in Australia and New Zealand. Growth is depressed at high biomasses by self-shading and competition for water and nutrients. When growth increment is converted to instantaneous growth rate the curve converts to a straight line of negative slope. Weak negative correlations were obtained between instantaneous growth rate and biomass for each component of the pasture in this study but only that for grasses on sandplain soil was statistically significant. Hence the relationship between growth and biomass appears to be much weaker in arid-zone pasture species than for sown species in more temperate environments. Rainfall modified by season of incidence, and the ephemeral strategy of evading drought, appear to be more important determinants of growth of pasture plants in the arid zone.

3.6   Conclusions

The effects of weather on soil moisture and on plant growth and biomass were assessed for the two major soil types on Kinchega National Park. Soil moisture was a function of rainfall and evaporation over the previous month. Instantaneous growth rate of pasture components was highly dynamic. Growth was best correlated with rainfall over the previous month or, in the case of perennial sub-shrubs on the sandplain soil, rainfall over the previous two months. Annual forbs and grasses grew and died back faster than did perennial sub-shrubs. Growth was initiated by a rainfall of more than 40 mm within a month and maintained by follow-up falls in succeeding months.

Biomass was related to rainfall over a longer time interval than growth (6 months or 12 months). Total vegetation biomass on both soil types could be predicted by rainfall over the previous 12 months. The season at which rain fell influenced the suite of plants that germinated.

4

Plant dynamics

GRAHAM ROBERTSON

4.1 Introduction

This chapter explores the following questions about vegetation in the sheep rangelands:

(1) How does rainfall affect plant biomass?

(2) How do rainfall and plant biomass affect plant growth?

(3) How does soil texture affect plant biomass?

(4) How does grazing by kangaroos and sheep affect plant biomass?

(5) Do any of these factors dominate to explain trends in plant biomass and species composition?

These questions will be addressed using data derived from Kinchega and Tandou. Pasture and shrubs are dealt with separately. Pasture includes all vascular plants in the herbaceous layer including chenopod sub-shrubs. Shrubs occupy the mid-layer of the vegetation and comprise perennial, deep-rooted, long-lived species relatively resistant to drought. The major shrub species is black bluebush (*Maireana pyramidata*), a 1.5 m high chenopod shrub not very palatable to kangaroos and sheep but eaten in times of severe food shortage.

4.2 Research methods

4.21 *Pasture measurements*

Pasture biomass and species composition were measured and species phenology was recorded at 213 sites on Kinchega National Park and at 100 sites on Tandou for each quarter between August 1980 and February 1984. Sites were located 1 km apart on parallel east-west lines separated by 2 km.

On each sampling occasion the pasture was measured at each site on three 0.25 m² circular plots positioned at random within the site. These plots were then caged to exclude grazing by kangaroos, sheep and rabbits and were remeasured three months later. Hence at each site on each sampling occasion I measured three plots for the first time and three plots that had been caged for three months. The new plots were then caged and the cycle repeated. Data from the three plots caged for three months were pooled at each site, as were the data from the three uncaged plots. From these measurements I sought on each occasion the biomass of the pasture grazed for three months and the biomass of the pasture protected from grazing for three months. Offtake by mammals, including the effect of trampling, was calculated by deducting grazed from ungrazed biomass at each site. Pasture growth or dieback was obtained by comparing grazed biomass at each site with ungrazed biomass at the same site three months later. All measurements were made in the presence of invertebrate herbivores. Their effects are unknown.

Pasture biomass was estimated by the comparative yield technique of Haydock & Shaw (1975). By this method reference plots were selected to span the range of biomass of each pasture type within the study area on each sampling occasion. The reference plots were photographed and clipped, the harvested vegetation being oven dried at 80°C and weighed. These series of photographs formed a standard reference point against which the biomass on sample plots was assessed visually.

Fig. 4.1. View from the air of the extensive bluebush communities that cover much of the sandplain of the study area.

Species composition was estimated by the dry-weight-rank method of Mannetje & Haydock (1963). It yields an estimate of botanical composition on a dry-weight basis without the need to cut and weigh samples of the pasture. Species in sampled plots were ranked by their contribution to biomass. A set of empirically derived multipliers (Mannetje & Haydock, 1963) converted these ranks to proportions which could then, if necessary, be converted further to kg/ha by multiplying them by total biomass estimated by the comparative yield method.

4.22   Shrub measurements

Black bluebush dominates the mid-layer vegetation on both Kinchega and Tandou. It occurs patchily or as continuous stands across much of the light-textured soils. In times of drought it is eaten by sheep and most of its foliage can be reached by kangaroos. Other potential browse species such as *Cassia* spp. and prickly wattle (*Acacia victoriae*) were not sampled either because of their limited occurrence or because most of their foliage is beyond the reach of the herbivores.

Offtake of bluebush by kangaroos on Kinchega and by sheep and kangaroos on Tandou was measured each six months between August 1982 and February 1984 by the movable cage technique (Brown, 1954; Pieper, 1978). Ten

Fig. 4.2. The bluebush community in Fig. 4.1 as viewed from the ground.

mammal-proof cages were placed around bluebushes at each of 20 sites on
Kinchega and at each of ten on Tandou. Offtake was measured as the
difference in the biomass of foliage between bushes that were protected from
browsing with those that were exposed to browsing. After each sampling the
cages were randomly relocated around another ten bluebushes. The biomass
of bluebush was estimated by the 'Adelaide technique' (Andrew, Noble &
Lange, 1979). Data were pooled at each site.

4.3   Pasture

4.31   *Trend over time*

Figure 4.3 shows the trend of pasture biomass on Kinchega and Tandou.
There were large differences among years, among corresponding seasons in
different years and among seasons in a given year. Differences between the
two study areas were minor. Analyses of variance detected small but significant
interactions between study area and season and between study area and year.
These differences probably reflect the variation in herbivore density between
Kinchega and Tandou during the study.

Above average rainfall in the autumn and winter of 1980 resulted in high
pasture biomass at the beginning of the study. Pasture biomass then dropped
by 50% on each of three consecutive sampling occasions. The first of these,
between August and November 1980, corresponded with post-flowering senescence and death of the annual forbs and grasses that dominated the pasture.

Fig. 4.3. Pasture biomass (± SE) on Kinchega National Park and Tandou
Property, 1980-84.

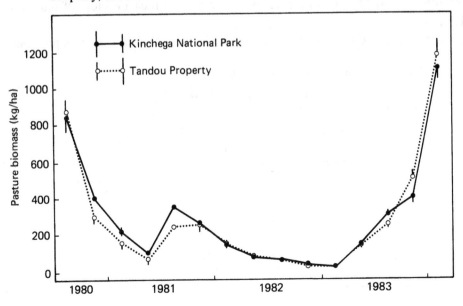

The second and third occurred during the summer of 1980-81. Kinchega received only one quarter of the long-term average summer rainfall during that summer and was very dry by May 1981. Some kangaroos were seen browsing bluebush and prickly wattle. Pasture was plentiful again over the winter of 1981 following above average rainfall. Menindee received less than half its long-term average rainfall in 1982 and under these conditions of drought the pasture biomass declined to 9 kg/ha on Kinchega and to 7 kg/ha on Tandou. Fourteen thousand kangaroos died on Kinchega between late-November 1982 and early-March, 1983 (Robertson, 1986, in press). The drought broke in March 1983 and pasture biomass responded to high rainfall during the remainder of 1983 and very high rainfall in early 1984.

The pasture biomass on both study areas changed by two orders of magnitude on two occasions in the three and a half years of the study. The second shift, during recovery from the drought, occurred in a single year. These rapid swings in biomass highlight the intrinsic lability of the system pastured by short-lived plants and subjected to erratic rainfall and to grazing. Menindee experiences severe drought about once every ten years and lesser droughts every three or four years.

4.32   Effect of rainfall

Studies of variation in plant productivity suggest that a large proportion of it can be accounted for by rainfall (Trumble, 1935; Lieth, 1973; Noy-Meir, 1973; Strugnell & Pigott, 1978). Pasture biomass on Kinchega was regressed on rainfall accumulated over the previous 1 month, 3 months, 6 months, 1 year, 2 years and 3 years before date of measurement of biomass (Table 4.1).

Rainfall accumulated over the previous six months and over the previous

Table 4.1. *Variance in pasture biomass accounted for by regression on rainfall over previous time intervals.*

Interval (months)	r^2
1	0.54
3	0.48
6	0.70
12	0.71
24	0.59
36	0.26

year were most closely associated with pasture biomass, accounting for 70% and 71% of variation respectively. Figure 4.4 shows the regression of biomass in kg/ha (y) on the previous six month's rainfall in mm (x):

$$y = -7.9 + 3.1x$$

which accounts for 70% of the variation and leaves a residual standard deviation of 178 kg/ha around the regression. A linear approximation is fitted, but as rainfall increases pasture biomass would eventually asymptote. The relationship of pasture biomass to rainfall did not differ between Kinchega and Tandou.

The relationship of pasture biomass to rainfall is affected by the species composition of the pasture. From 1980 to 1983 the pastures on Kinchega and Tandou were dominated by annual forbs and grasses. These have rapid acceleration and decline phases and a 2-4 month growing season. Biomass of annual plants in the pasture peaked every 6 months or so if rain fell in both summer and winter, and every year if rain fell in just one of these seasons. In contrast, the biomass of perennial plants in the pasture was more closely linked to short-term rainfall; rainfall over the previous month explained 89% of variation. Unlike annuals that have to grow from seed when conditions are favourable, perennials have an established root system and are able to utilise falls of rain more rapidly.

Fig. 4.4. Regression of pasture biomass (Kinchega National Park) on the rainfall for the 6 months before each biomass estimate.

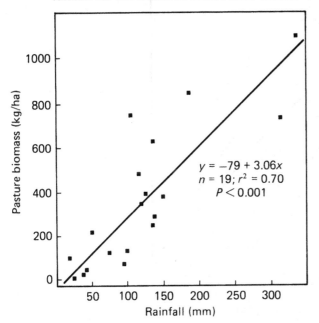

$y = -79 + 3.06x$
$n = 19; r^2 = 0.70$
$P < 0.001$

Blaisdell (1958) reported that grass and forb production in eastern Idaho increased by 1.7 kg/ha for each 1 mm increase in rainfall. In a desert area of southern Idaho total vegetation productivity increased by 5.6 kg/ha for every 1 mm increase in artificial precipitation (Pearson, 1965a). The productivity of total vegetation in desert communities in Namibia increased by 10 kg/ha for each 1 mm increase in annual precipitation (Pearson, 1965b citing Walter, 1955). In these studies the vegetation was protected from grazing. On Kinchega the biomass of pasture plants exposed to grazing increased by 3 kg/ha for every 1 mm increase in rainfall.

Rainfall predicted plant growth better than it predicted pasture biomass. However, the growth of plants is affected by their biomass. Plant growth tends to slow as competition for water and nutrients increases with rising plant density (Harper, 1977). The commonest trend of growth increment on plant biomass approximates a parabola (Noy-Meir, 1975). The relationship was examined by fitting a parabola to the growth increment of pasture protected from grazing over three months against the pasture biomass at the beginning of each period. Rainfall over the three months of growth was added in as another independent variable. The fitted regression of three-month

Fig. 4.5. Plant growth over three months as a function of plant biomass at the beginning of the interval and rainfall during the interval, for pastures on Kinchega National Park.

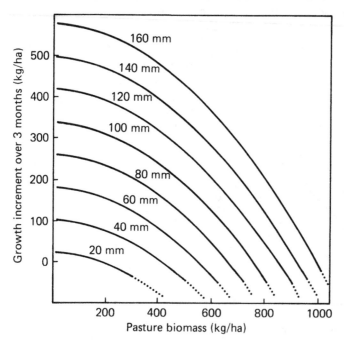

growth increment in kg/ha (*Y*) was related to rainfall in mm over the same three months (*R*) and starting biomass (*X*) by
$$Y = -55.12 - 0.01535X - 0.00056X^2 + 3.946R$$
That regression accounts for 97.3% of the variance in pasture growth over successive three-month periods between August 1980 and February 1984 ($n=$ 14) and leaves a residual standard deviation of 52 kg/ha around the regression. Of the two independent variables, starting biomass and rainfall, the latter is the more powerful predictor, accounting for 83% of variance when growth is regressed on it alone.

Figure 4.5 shows the relationship between growth over three months, starting biomass and rainfall. The parabolic effect of starting biomass has no descending left limb. A descending left limb has been reported consistently for plant growth measured over short intervals and its absence here is almost certainly an artefact of the long intervals between the measurement of initial and final biomass in the caged plots. The full parabola would be expected if growth had been measured daily.

Fig. 4.6. The isocline of zero growth for pasture on Kinchega National Park in terms of rainfall over three months and pasture biomass at the beginning of those three months.

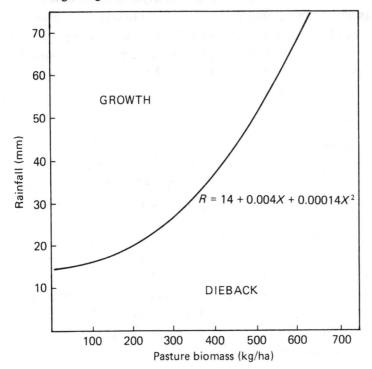

The equation above may be rewritten to estimate the rainfall over three months resulting in no pasture growth:

$$R = 14 + 0.004X + 0.00014X^2$$

which is the isocline for zero growth (Fig. 4.6). Thus if pasture biomass is zero a rainfall of more than 14 mm is necessary to trigger growth (in this case germination). With increasing pasture biomass the rainfall needed to produce further growth increases such that at 500 kg/ha more than 51 mm is needed and at 1,000 kg/ha more than 158 mm is required.

4.33   Effect of soil texture
Pasture biomass
There was no overall difference in pasture biomass between heavy-textured and light-textured soils on Kinchega.

Growth and dieback
Over the study period Tandou received nearly 100 mm more rainfall than Kinchega. Most of this extra rain fell during the winter and spring of 1981. Hence some growth responses on Tandou (e.g. in November, 1981) were greater than those on Kinchega. Overall, growth and dieback of pastures differed little between study areas (Fig. 4.7). Analysis of variance revealed that growth and dieback varied markedly between years and between the corresponding seasons of consecutive years in response to erratic rainfall.

There were minor differences in growth and dieback between heavy-textured and light-textured soils but overall the pattern of response was similar. Differences emerged during the summer of 1981-82 when rainfall triggered growth only on light-textured soils.

4.4   Effect of grazing
The trends of sheep and kangaroo densities on Tandou and of kangaroo density on Kinchega are shown in Fig. 4.8. To allow a rough comparison the sheep are converted to kangaroo equivalents by multiplying their density by 1.6 according to their food intake (an average of estimates by Foot & Romberg, 1965; Griffiths & Barker, 1966; McIntosh, 1966; Forbes & Tribe, 1970; Hume, 1974; Chapter 6). The differences in kangaroo equivalents between study areas were probably less than those shown. Sheep on Tandou were moved often and sometimes they did not graze the study area for the full period between samples. Because sheep outnumbered kangaroos on Tandou, and because they individually eat more than kangaroos, they consumed about 70% of the total offtake on Tandou.

The density of kangaroos on Kinchega varied as a result of births and deaths whereas that of the sheep on Tandou reflected management decisions to buy, sell and shift stock onto and off the area I studied. For example, many sheep were mustered off the study area and onto the dry bed of Tandou lake over the summer of 1981-82 because of a dry spell. Sheep were moved off the

Fig. 4.7. Growth and dieback (± SE) of pastures on Kinchega National Park and Tandou Property. Estimates are for 3-month intervals.

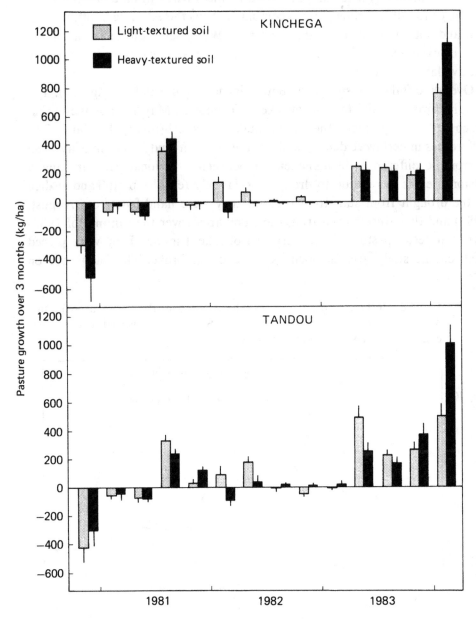

study area in January 1983 because of the drought and were again removed in January 1984 because of flooding from the Darling River. Kangaroos on Tandou ranged as freely as those on Kinchega but were occasionally harvested, sometimes heavily, for the pet food market.

The method used to determine offtake may result in estimates that are too high, particularly during periods of rapid plant growth (Sharrow & Motazedian, 1983). However, analysis of variance revealed a significant difference between grazed and ungrazed biomass (the estimate of offtake) on Kinchega for all sampling periods combined. Thus the estimate of offtake was real and not an artefact of sampling error. While some estimates of offtake may be exaggerated, the overall trend in offtake was qualitatively similar to the availability of pasture.

Overall, offtake was similar on both Kinchega and Tandou (Fig. 4.9). The high herbivore density but low offtake on Tandou in May 1981 suggests that sheep had either reduced their food intake or were browsing bluebush. The differences in herbivore density and offtake between study areas in May 1983 mirror the differences in management between a national park and many pastoral leases in relation to drought. Relatively few sheep on Tandou died in the drought from starvation because they were hand-fed (from August, 1982) and they were moved from the study area (over the summer of 1982-83) onto better pastures in the dry bed of lake Tandou. They were placed back on the study area as soon as the drought broke. The kangaroos on

Fig. 4.8. Herbivore densities on Kinchega National Park (kangaroos) and Tandou Property (sheep and kangaroos). Sheep are converted to kangaroo equivalents according to their food intake.

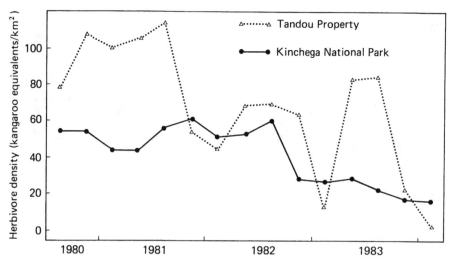

Kinchega were not buffered against the effects of the drought. About 50% of them died (Chapter 8).

Grazing had the highest relative effect on pasture, but the lowest absolute effect, during the drought. In a mammal-proof exclosure on Kinchega designed to determine the pasture biomass in the absence of grazing, biomass was 167 kg/ha in August 1982 when the drought began to deepen compared to 9 kg/ha in a similar area grazed heavily by kangaroos and rabbits. When the drought peaked, pasture biomass in the exclosure yielded 113 kg/ha but only 11 kg/ha under grazing and trampling. At the peak of the drought pastures across Kinchega stood at only 9 kg/ha and offtake by kangaroos declined to about one-tenth of the average for the study.

The offtake from pastures on different soil textures is best examined on Kinchega where kangaroos range freely (Fig. 4.10). Between November 1981 and May 1982 plants grew only on light-textured soils and kangaroos removed approximately four times as much forage from them as from heavy-textured soils. Offtake was again higher on light-textured than on heavy-textured soils in November 1982 following light rainfall in September of that year.

Most of the sheep rangelands are grazed by rabbits which build warrens only in light-textured soils. They are absent from the heavy-textured soils of

Fig. 4.9. Pasture biomass (± SE) removed by herbivores (grazing and trampling) from Kinchega National Park and Tandou Property. Estimates are for 3-month intervals.

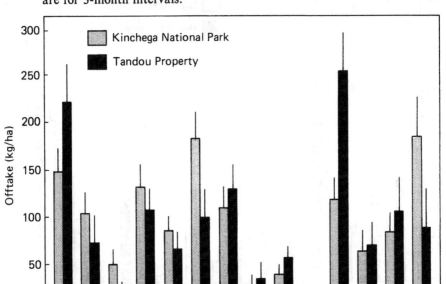

the floodplain of Kinchega and Tandou. Rabbit densities fluctuated with weather events and disease. They were abundant on Kinchega in the mid-1970's but drought in 1977 and a heavy outbreak of myxomatosis in 1978 reduced their density to extremely low levels at the commencement of the study. Mild outbreaks of myxomatosis in 1981 and 1982, the drought of 1982-83, and control by park managers kept rabbit densities low throughout the study. Spotlight counts of rabbits conducted quarterly on Kinchega showed that rabbit numbers peaked during this study late in 1981 when one rabbit was counted every 5 ha. There were fewer rabbits than this on Tandou. Because rabbits on Kinchega were outnumbered by kangaroos and because a rabbit eats about one-tenth as much forage as a kangaroo (Chapter 6), they were of little consequence as grazers during the study.

4.5   Pasture budget

Pasture biomass fluctuated in response to growth, dieback and offtake by herbivores. While these features have been treated separately the relative contribution of each to pasture dynamics can be summarised by

Fig. 4.10. Pasture biomass (± SE) removed by kangaroos from light-textured soils and heavy-textured soils on Kinchega National Park. Estimates are for 3-month intervals.

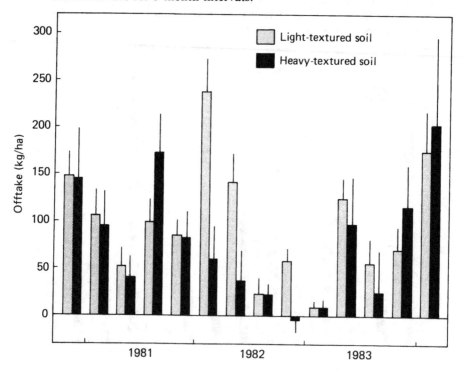

Fig. 4.11. Pasture budgets for Kinchega National Park and Tandou Property. Changes in pasture biomass are expressed as a function of plant growth, dieback and offtake by herbivores. Estimates are for 3-month intervals.

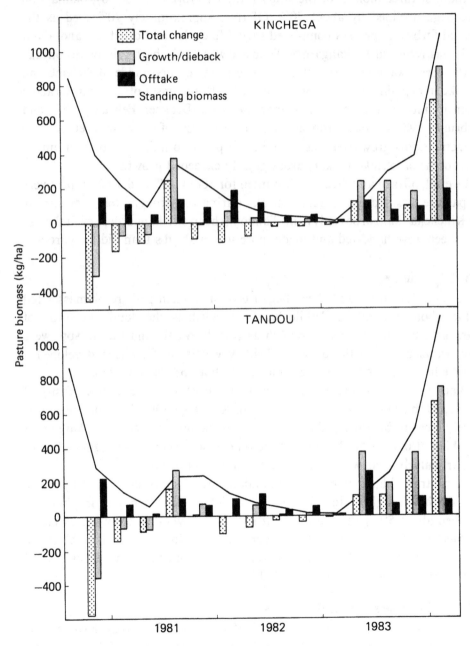

combining them as a pasture budget (Fig. 4.11). The trend in pasture biomass, growth, dieback and offtake was similar for Kinchega and Tandou. When pasture biomass was high, dieback reduced it more than offtake. For example, in the first three months of the study (August-November, 1980), biomass on Kinchega declined by about 50%. Of this, approximately 70% died as the annual forbs and grasses completed their life cycles and dried off, and about 30% was removed by kangaroos. Dieback was again high relative to offtake in the corresponding season of 1981. The relative importance of dieback and offtake in explaining pasture loss changed with the onset of drought. Grazing intensity increased as pasture biomass declined. Between February 1982 and February 1983 biomass dropped by over 90%, most of which was removed by grazers. Plants grew over this period in response to light falls of rain but on each occasion offtake by kangaroos greatly exceeded growth.

Overall, offtake and dieback accounted for about 70% and 30% respectively of pasture loss. Rate of pasture offtake during the drought was very similar on Kinchega (kangaroos only) and Tandou (sheep and kangaroos) even though the sheep were hand-fed and made more use of shrubs than did kangaroos.

4.6   Shrubs

Kangaroos and sheep began browsing when pasture biomasses fell below about 100 kg/ha. While black bluebush is the dominant shrub of browsable height, other shrubs such as prickly wattle and *Cassia* spp. were also browsed during the drought. Prickly wattle on Kinchega developed a browse line consistent with the removal of about 5% of the foliage.

Black bluebush is long-lived and drought resistant. It reaches a density of over 2,000 bushes/ha across about 20% of Kinchega. Offtake of bluebush was barely measurable during the drought even though the density of kangaroos was high at 0.5 per ha. Kangaroos ate bluebush during the drought (Chapter 5) particularly near water, but they could not survive on it.

In contrast, black bluebush on Tandou was severely defoliated by sheep during the drought. Most bushes were defoliated when pasture biomass reached about 50 kg/ha. However, it is doubtful whether browsing by either herbivore is detrimental to black bluebush in the long-term. Black bluebush is probably more widespread now than before settlement in areas grazed continually by sheep (Barker, 1972, 1979).

4.7   Changes in species composition

Winter and summer rains differ in the effect they have on the species composition of the pasture at Menindee. Winter rainfall encourages the germination and growth of annual and perennial forbs, annual grasses and perennial sub-shrubs. Moderate summer rainfall encourages perennial forbs, annual grasses and perennial sub-shrubs. While there is a wide seasonal overlap of plant groups many species grow only in winter or in summer. Superimposed on this cycle is an episodic change in species composition occasioned by heavy summer rainfall. The pasture erupts in a profusion of vegetation comprising species that have lain dormant in the seed bank, often for several years, until reactivated by this super-stimulus. These rainfall events produce pastures dominated by perennial sub-shrubs, perennial grasses and perennial forbs. Such rains fell at Menindee over the summer of 1983-84.

The perennial grasses that grow with high summer rainfall are short-lived. They have a life history strategy that differs from that of the perennials of a temperate grassland. They become dominant quickly but relinquish their dominance after a year or two. For example, Williams & Roe (1975) reported that 96% of *Stipa variabilis* plants died within two years of emergence and none survived to three years. Similarly, the life span of *Eragrostis setifolia* rarely exceeds three years. The sheep rangelands do contain long-lived graminoids (*Danthonia caespitosa* in the Riverine Plain of New South Wales and Mitchell grass, *Astrebla* spp., of the drier tropics and subtropics) but they do not occur commonly at Menindee.

Figure 4.12 traces the composition of the pasture over the course of the study. It shows the rapid acceleration and decline phases of annual plants and the effect of heavy summer rainfall. Overall, perennial sub-shrubs were the most stable component of the pasture. They persisted through all seasons but their contribution to biomass was usually lower than that of the annual forbs and grasses. In the peak of the drought perennial sub-shrubs comprised approximately 70% of pasture biomass.

Over the summer of 1983-84 Kinchega and Tandou received twice the average rainfall for summer, including 131 mm in one month (January). Plants were large and profusely reproductive on both study sites. Pasture was dominated by perennial sub-shrubs, principally *Sclerolaena* spp. and *Atriplex* spp. and by perennial grasses such as *Chloris truncata, Eragrostis dielsii, E. setifolia, Stipa variabilis* and *Triraphis mollis*. Ninety-nine species were represented in the pasture, perennials constituting 67% of the biomass. *Enneapogon avenaceus*, an annual grass that behaves as a perennial given suitable conditions, was the major dominant and survived throughout 1984. In contrast, during the peak of the drought the pasture comprised only ten species of which eight were woody sub-shrubs. The presence of perennial grasses at the commencement of the study is testimony to the high rainfall (170 mm) over the summer of 1978-79.

The resurgence of perennials was triggered by heavy summer rain. Most

Fig. 4.12. Trends in species composition of pasture on Kinchega National Park (grazed by kangaroos) and Tandou Property (grazed by sheep and kangaroos).

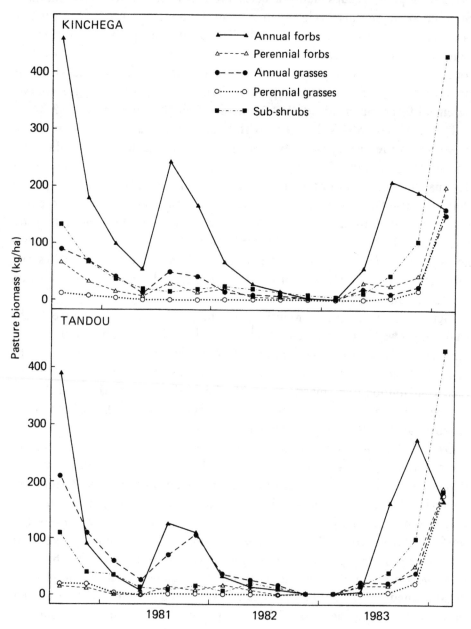

plants grew from seeds. Over the summer of 1983-84 Kinchega received 210 mm of rain, almost twice the long-term average for summer. Falls of similar timing and intensity have been recorded about once every seven years over the last 45 years at Menindee. Summer rainfall in excess of 200 mm is probably a sufficient trigger. It occurs on average every five years.

In the summer of 1984-85 pasture biomass did not decline as it did in the summers of 1980-81 and 1981-82. Instead it remained high, between 400 and 700 kg/ha. Although weather conditions over the winter of 1984 favoured the growth of annuals, perennials continued to dominate the pasture presumably by inhibiting the growth of annual seedlings. Many of the forbs that managed to establish were stunted at maturity. Even allowing for reduced post-drought grazing pressure by mammals it seems that perennials dampened the fluctuations in biomass.

The changes in species composition recorded here suggest that the pastures at Menindee swing in dominance between annuals and perennials. These changes appear to be influenced more by heavy rainfall than by grazing. Changes in species composition of *Astrebla* grassland associated with high rainfall in south-western Queensland led Orr (1981) to a similar conclusion.

Although the species composition of the pasture may change, the regime of moisture and grazing at Menindee should, overall, favour the growth characteristics and reproductive strategies of annuals more than perennials. Annuals grow only when conditions are favourable and they usually over-summer as seed. They may tiller more rapidly than perennials (Silcock, 1977) and be phenotypically more plastic (Young, Evans & Major, 1972). They may be more competitive for light than perennials (Silcock, 1980). Annuals have large banks of viable seed (Westoby, 1979-80) that are unlikely to become depleted (Mott, 1972) and they have germination requirements suited to episodic and unreliable rainfall (Trevis, 1958; Cohen, 1966). Annuals can set seed despite periodic adverse weather and stocking conditions (Williams, 1961). Continuous grazing favours them by reducing perennials and thereby reducing competition. Perennials are grazed particularly during summer when winter annuals are senescent or dead. Annuals are less likely to be affected by grazing because they grow mostly in winter when the pastures are usually widespread. Australian pastoralists usually prefer a high proportion of annuals in the pasture because these are highly palatable to sheep (Lorimer, 1978; Orr, 1978).

4.8   Conclusions

Rainfall and grazing by mammalian herbivores are the major determinants of pasture biomass in this region of the Australian sheep rangelands. The effects of extreme weather events, such as high rainfall and drought, overshadow the effects of grazing. When the pasture is dominated by annual plants rainfall over six months explains 70% of the variation in biomass over that period. Rainfall over three months, winter or summer, predicts the growth increment of pasture over that interval. The prediction is tightened a little by including within the equation a term for pasture biomass at the beginning of the interval.

Biomass and species composition of pastures grazed by kangaroos alone were similar to those grazed by a mix of kangaroos and sheep. The similarity may reflect both study areas having been grazed by sheep and rabbits for many decades prior to 1970. Grazing by mammalian herbivores accounted for about 70% of the pasture biomass removed on both areas.

Kangaroos differ from sheep in the effect they have on bluebush shrublands. Sheep browse bluebush heavily during droughts. Kangaroos eat it sparingly, even when pasture is depleted, and then usually only near water. However, there is no evidence from other studies that browsing by either herbivore has any long-term effect on black bluebush.

5

The diet of herbivores in the sheep rangelands

R. D. BARKER

The diets of herbivores have been studied by many people throughout the arid zone of Australia. Here their findings are synthesised and compared with those from red and western grey kangaroos on Kinchega National Park. Finally the role of kangaroos in the dispersal of seeds is examined.

5.1 Methods

The investigation of diet on Kinchega began in November 1980 and ended in February 1984. Plant species were collected throughout the park and the specimens milled to make reference slides. Diets were assessed periodically from faeces collected in the field ('field faeces' hereafter) and from the stomach contents and rectal contents of animals shot to provide data on condition and reproduction (Chapter 9).

The field faeces were collected only from animals that were seen to void and so there is no possibility of the faeces being ascribed to the wrong species. Caughley (1964) and Grant (1974) maintained that the kangaroo species producing a faecal sample could be determined from the size and shape of the pellets but I was often unable to identify pellets to species, particularly after rain had produced a lush pasture. In such conditions the faecal material maintained no characteristic shape.

All samples of plant, stomach and faecal material were prepared by drying, grinding, decolourising with ethyl alcohol and commercial bleach, washing and mounting on slides in glycerine gel. These samples of stomach contents and faecal material were scored by their content of grass, forbs, shrubs (excepting prickly wattle and chenopod bluebush), prickly wattle (*Acacia victoriae*), chenopod sub-shrub and chenopod bluebush.

For the purposes of this analysis the plants of the family Chenopodiaceae (collectively referred to as chenopods) are divided into two groups according to their growth habit. 'Chenopod sub-shrubs', mainly species of *Sclerolaena* (=*Bassia*) and *Atriplex*, may be annual or perennial, are shallowly rooted, and form part of the pasture layer. They react to a shower of rain almost as quickly as grasses and forbs. In contrast, 'chenopod bluebushes' (*Maireana* spp.) are long-lived perennials. They grow 1-2 m high and may have a crown diameter of 2 m. Being rooted deeply they withstand the most savage of droughts. Nine species of bluebush are recorded from Kinchega National Park but the most common is black bluebush (*Maireana pyramidata*).

It will be shown elsewhere that the technique of faecal analysis produces inaccurate results. It is not possible to reconstruct the diet accurately by counting plant fragments in faeces and apportioning them among species. Plant species differ in digestibility, in the number of fragments that they provide per unit of dry weight, and in the proportion of those fragments that can be identified. Attempts to produce correction factors have to date been rewarded with limited success (Barker, 1986 a). No such attempt is made here. The amount of plant material on a slide ascribable to a plant species or species group was scored simply as none (0), a little (1), some (2) or a lot (3) (Barker, 1986 b). These scores were then averaged across replicates.

Seeds were counted in rectal faeces of specimens shot at five sites across the park in February 1981. Ten cubic centimetres (by displacement) were teased apart and all seeds identified and counted. Seeds were counted also in faecal samples collected at the end of an experiment (Short, 1986, and unpubl. data) examining the grazing efficiencies of red kangaroos, grey kangaroos and sheep in September 1983. The animals were penned in adjacent 20 x 20 m yards and allowed to graze the vegetation to the ground. The samples were passed through a series of sieves of decreasing aperture following the technique of Wells (1973) to determine how finely each species ground its food.

5.2   Results

5.21   *Dietary differences over seasons*

Figures 5.1 and 5.2 indicate that although there were large changes in diet over the years 1981 to 1984 these were not seasonal effects. There are differences between the two species of kangaroo but within a species the proportion of the various plant categories in the diet was relatively constant across spring, summer, autumn and winter pooled over years.

5.22   *Dietary differences over time*

Figure 5.2 gives the trend of grasses, forbs, chenopod sub-shrubs, shrubs, prickly wattle and bluebush in the diet of the two species between

1981 and 1984. It is most easily interpreted by reference to Fig. 4.11 which shows the pasture budget and Fig. 4.12 which shows the species composition of the pasture over the same period.

Red kangaroos

The red kangaroos ate a range of plants including grasses, forbs, chenopod

Fig. 5.1. The mean of the scoring of the amount of grasses, forbs, chenopods and shrubs in the diet of red and western grey kangaroos for each season at Kinchega National Park for the years 1981-4. Su – Summer, Au – Autumn, Wi – Winter, Sp – Spring.

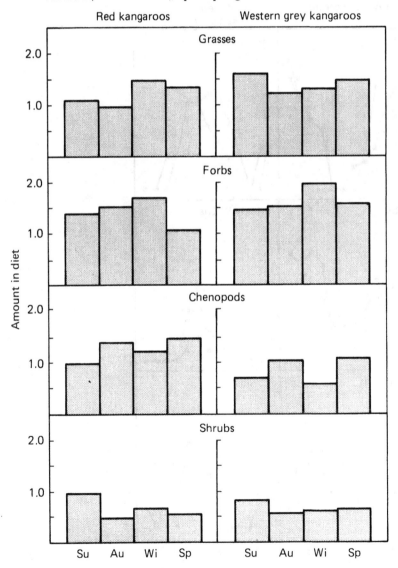

sub-shrubs and some shrubs following the dieback of vegetation over the summer and autumn of 1980-81. The winter rains of 1981 (138 mm) produced enough moisture for a flush of growth in spring of which the reds took advantage. They ate grasses (mainly *Lophochloa pumila*) and forbs (including *Senecio* spp., *Myriocephalus stuartii* and various species of Malvaceae and

Fig. 5.2. The change in the amount of grasses, forbs, chenopods, shrubs, prickly wattle and bluebush in the diets of red and western grey kangaroos at Kinchega National Park, 1981-1984. The vertical axis numbers of 1, 2 and 3 refer to a little, some, and a lot, respectively.

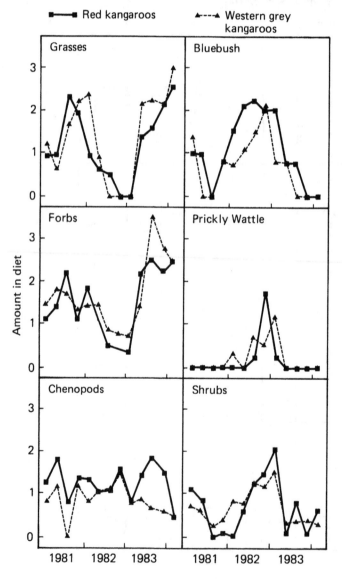

Solanaceae). There were virtually no shrubs, bluebush or prickly wattle in the diet and very little chenopod sub-shrub.

There was scant rain for the rest of the year and, as the grasses and forbs senesced, the reds shifted to a diet of chenopod sub-shrubs which they maintained until the beginning of 1982. Eighty-two millimetres of rain fell in the first three months of 1982 and this promoted a flush of winter-growing plants including the forb *Plantago drummondii* and species of Malvaceae and Solanaceae which the reds ate. The flush was quickly eaten out and the reds' intake of grasses and forbs declined through to the end of 1982. The lack of winter rain and the light spring rains of 1982 were enough to allow the *Sclerolaena* spp. to put on new growth and these provided much of the food of the reds until the end of the year.

Shrubs formed an increasing proportion of the red kangaroos' diet as the grasses and forbs declined in biomass through 1982. Bluebush was eaten increasingly and prickly wattle entered the diet in significant proportions in November 1982. Shrubs known to be eaten included species of *Cassia* and *Eremophila*. At the peak of the drought in February 1983 the reds ate shrubs and some chenopods (mainly *Chenopodium nitrariaceum* and bluebush).

Fig. 5.3. An area of red sandy soil at the peak of the drought in February 1983. There are several small bluebushes in the foreground, two dead prickly wattle trees in the immediate background and hopbushes and emu bushes in the distance. The dark spots on the ground are kangaroo faeces.

The drought was broken by a fall of 62 mm in March 1983 followed by 33 mm in May. The annual grasses and forbs responded almost immediately. The reds ate grasses such as *Lophochloa pumila* and *Enneapogon avenaceus*, forbs such as *Boerhavia diffusa*, and various species of Malvaceae and Solanaceae, Brassicaceae, Amaranthaceae (including *Glinus lotoides* and *Alternanthera* spp.), Asteraceae (including *Senecio* spp., *Centaurea melitensis* and *Myriocephalus stuartii*) and medics (*Medicago* spp.). Reciprocally the intake of shrubs and bluebush in the diet fell to a low level.

Western grey kangaroos

The first samples collected in February 1981 showed that the western greys were eating grasses (*Enneapogon avenaceus*), forbs (mainly *Glinus lotoides* and Brassicaceae species), a little bluebush and shrubs. In September 1981, when winter rains had allowed a flush of new growth, the western greys increased their intake of grasses (mainly *Lophochloa pumila, Agrostis aven-*

Fig. 5.4. The same area as shown in Fig. 5.3 in September 1983 after the breaking of the drought. The ground cover is mainly of saltbushes (*Atriplex spp.*) and bottlewashers (*Enneapogon avenaceus*).

acea and *Hordeum leporinum*) and forbs (including *Plantago drummondii* and species of Malvaceae, Solanaceae and Asteraceae) at the expense of shrubs and chenopod sub-shrubs. From a peak in February 1982 the intake of grass declined steadily until by September none was eaten. Forbs were still being eaten in that month (*Plantago drummondii* and species of Malvaceae, Asteraceae and Brassicaceae) but these also had dropped out of the diet by February 1983.

Chenopod sub-shrubs put on new growth with the light spring rains of 1982 and were eaten by the grey kangaroos at that time, but from then on their contribution to intake declined. Bluebush was eaten increasingly to September 1982 but declined in the diet thereafter. The contribution of shrubs to the diet (particularly prickly wattle and lesser quantities of *Dodonaea attenuata* and *Eremophila* spp.) increased throughout 1982 to a peak in February 1983.

With the breaking of the drought in March and May 1983 the greys began to eat the newly-grown grasses and forbs and these formed the bulk of the diet thereafter.

5.23   *Kangaroos as dispersers of seed*

Table 5.1 records the prevalence of seeds in the faeces of kangaroos. Analyses of variance detected no difference between species or sexes of kangaroos in the number of seeds in the faeces. However when the data were analysed as the number of species of plants represented by those seeds a significant difference emerged between sexes. The faeces of female kangaroos carried the seeds of more plant species than did those of the males.

The results of the sieving experiment are shown in Fig. 5.5. Neither species of kangaroo grinds its food as finely as do sheep. Thus a seed ingested by a kangaroo has a greater chance of being excreted whole and hence viable than a seed ingested by a sheep.

Table 5.1. *Number of seeds in 10 cc (by displacement) of faecal pellets of red and western grey kangaroos collected from five localities on Kinchega National Park in February 1981. The numbers of species of plants in each sample are shown in brackets.*

Kangaroo species	Locality				
	1	2	3	4	5
Red male	234(9)	136(11)	66(8)	225(12)	600(9)
Red female	241(11)	171(10)	219(9)	209(15)	447(14)
Grey male	215(10)	149(9)	56(5)	797(11)	496(10)
Grey female	267(11)	340(13)	96(10)	449(11)	292(10)

The number of seeds in the faecal samples from the graze-down experiment are shown in Table 5.2. The faeces of sheep contained significantly fewer seeds and significantly fewer species were represented by those seeds than in the faeces of either kangaroo species.

5.3   Discussion

Throughout the study the diet of kangaroos was affected by the food on offer. When grasses and forbs were present these were taken. As pasture conditions deteriorated the animals were forced to utilise chenopod sub-shrubs and shrubs. Generally they took the better quality species of plants present at any time.

The plants themselves reacted strongly to the weather. Depending on the timing of rainfall a single genus or species of plant will often dominate a pasture. One area of the park was dominantly vegetated by a sward of *Sclerolaena* spp. at the beginning of the study. This was succeeded by *Angianthus brachypappus* and later by *Babbagia acroptera*. Another area was dominantly carpeted successively by *Hordeum leporinum*, *Plantago drummondii* and *Agrostis avenacea*. As western grey kangaroos and most red kangaroos are comparatively sedentary (home range ⟨8 km², Chapter 7) they tend to utilise these areas. These sedentary kangaroos do not move even when

Fig. 5.5. The percentage weight of faecal material of reds, western greys and sheep retained by sieves of decreasing aperture. R represents the material which passed all sieves.

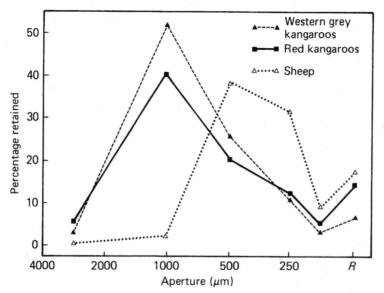

food supplies are severely depleted and if this source fails they become too weak to move and die.

The vegetation present in any area was a function of rainfall which was effective on only two occasions during the study: the spring of 1981 and the break of drought in March-May 1983 and subsequent falls during the rest of the year. The high biomass present at the beginning of the study was the result of good winter rains in 1980. The red kangaroos began eating grasses before the greys in the spring of 1981, the reverse occurring in autumn 1983. The main species of grasses eaten by both species of kangaroos were *Enneapogon avenaceus*, *Eragrostis* spp. and *Lophochloa pumila*. Additionally the greys ate *Stipa* spp. and *Agrostis avenacea*.

Half of the samples from red kangaroos and a third of those from the greys contained species of Malvaceae (mainly *Sida* spp.) and Solanaceae (*Solanum esuriale* and *S. karsensis*). Other plants eaten in approximately the same amounts by both species were various species of the families Asteraceae (including *Senecio* spp., *Myriocephalus stuartii*, *Centipeda cunninghamii* and *Centaurea melitensis*), Brassicaceae (including *Sisymbrium* spp. and *Brassica tournefortii*), Plantaginaceae (*Plantago drummondii*) and Euphorbiaceae

Table 5.2. *The number of seeds of different species of plants occurring in duplicate faecal samples (10 cc by displacement) of red and western grey kangaroos and sheep grazing in adjacent pens on Kinchega National Park in September 1983.*

Plant species	Red kangaroo		Western grey kangaroo		Sheep	
Euphorbiaceae	38	11	102	52	72	34
Bulbine bulbosa	23	5	2	21	0	0
Poaceae 'a'	18	7	4	5	5	0
Enneapogon avenaceus	16	14	0	1	5	6
Poaceae 'b'	13	6	4	75	20	22
Amaranthaceae	10	9	0	0	3	2
Sclerolaena sp.	10	0	1	1	1	0
Brassicaceae	7	1	0	0	1	0
Boerhavia diffusa	6	1	1	3	0	0
Lophochloa cristata	1	7	0	0	0	5
Eragrostis sp.	0	0	3	14	0	0
Plantago drummondii	0	5	34	50	0	0
Other species	13	5	10	8	1	0
Total species present	16	13	14	14	8	5

(including *Phyllanthus* spp.). Twice as many samples from greys as from reds included plants of the Leguminosae (including *Medicago* spp., *Trigonella suavissima* and *Swainsona* spp.). Overall forbs were eaten more by greys than reds.

Chenopods other than bluebush were eaten consistently by both species of kangaroo throughout the study period except after the winter and spring rains of 1981. The pasture during the September 1981 sampling period was dominated by a flush of forbs and grasses and the grey kangaroos had no chenopod sub-shrubs or bluebush in their diet at that time. The very light winter rains of 1982 were insufficient to repeat the previous year's flush of grasses and forbs but were enough to trigger growth of some *Sclerolaena* species. These were eaten by both species of kangaroo. Generally the reds ate more chenopod sub-shrubs than the greys throughout the entire study period.

Bluebush was similarly eaten more by red kangaroos than by greys. It had became a major component of the diet of reds by May 1982 and remained as such until the drought broke ten months later.

The red kangaroos initially ate more shrubs (mainly *Cassia* spp., *Eremophila* spp. and *Dodonaea attenuata*) than the greys but this behaviour reversed after September 1981. At the peak of the drought the reds' intake of shrubs was a little higher than the greys'. Prickly wattle was eaten throughout 1982 by grey kangaroos, the intake peaking in February 1983. The reds did not start eating it until September 1982.

Table 5.3 summarises the findings of previous studies on the diet of grazing animals in the arid zone of Australia. It shows that red kangaroos prefer to eat grasses and forbs but when conditions deteriorate they switch to chenopods and in some areas will even browse shrubs. Eastern grey kangaroos (*M. giganteus*) eat more grass than do reds and will continue to search for it as long as it is available. Euros (*M. robustus*) also prefer grass but occasionally eat shrubs. The only previous information on the diet of the western grey is Curtis's (1973) comment that it eats less grass and more shrubs than the eastern grey or the red.

Sheep prefer grasses and forbs but shift more readily to shrubs, chenopod sub-shrubs and dry material than do kangaroos. Cattle are less selective of grass species than are kangaroos and will eat shrubs during drought. The only study of the diet of rabbits (*Oryctolagus cuniculus*) in the arid zone is that of Dawson and Ellis (1979). Rabbits eat grasses and forbs in times of plenty and chenopods and shrubs during drought.

Changes in the relative amounts of grasses, forbs, chenopod sub-shrubs, chenopod shrubs and shrubs from times of plenty to those of drought are shown in Fig. 5.6 for sheep, red, eastern and western grey kangaroos in

Table 5.3. *Diet studies on mammalian herbivores in the arid regions of Australia (after Squires, 1982).*

Animals	Location	Diet summary	Authors
Cattle and red kangaroo	Alice Springs	Both ate grasses; kangaroos preferred *Eragrostis setifolia* and forbs; cattle preferred shrubs and chenopods.	Chippendale, 1962
Cattle	Alice Springs	Preferred grasses and forbs but use browse as backup in dry times.	Chippendale, 1964, 1968a
Eastern grey kangaroos	Warwick Qld	Diet mainly grasses but no selection in comparison with pasture.	Kirkpatrick, 1965
Sheep	Deniliquin N.S.W.	Sheep selected grasses and forbs, then saltbush, dry material and medic burr (bladder saltbush/cotton bush vegetation association).	Leigh & Mulham, 1966a; Wilson *et al.*, 1969
Sheep	Deniliquin N.S.W.	Grasses and forbs when available, at other times bladder saltbush. Pigface not eaten (bladder saltbush/pigface association).	Leigh & Mulham, 1967
Sheep	Deniliquin N.S.W.	Forbs and perennial grass *Danthonia caespitosa* eaten in spring and winter. *Chloris truncata* and *Sporobolus caroli* selected in summer; small amount of cotton bush (cotton bush/grassland association).	Leigh & Mulham, 1966b; Leigh *et al.*, 1968
Sheep	Deniliquin N.S.W.	Annual grasses and forbs in spring *Danthonia caespitosa* at other times medic burr increases in the diet as *Danthonia* decreases (grassland).	Robards *et al.*, 1967
Sheep, eastern grey & red kangaroos	Cunnamulla Qld	Greys ate more grass and less forbs than reds. Sheep ate more browse than the kangaroos but overall diet was similar to the reds.	Griffiths & Barker, 1966
Euros	Pilbara W.A.	Grasses were the main part of diet; spinifex spp. were the main grasses eaten.	Ealey & Main, 1967
Sheep, euros & red kangaroos	Pilbara W.A.	All species ate grasses; kangaroos & sheep had similar diets; euros ate more soft spinifex; *Aristida* spp. eaten by macropods but not by sheep.	Storr, 1968
Red kangaroos	Alice Springs N.T.	Grasses (mainly *Eragrostis setifolia*) most important part of diet; forbs in spring and a little browse in summer.	Chippendale, 1968b

Table 5.3. *Diet studies on mammalian herbivores in the arid regions of Australia (after Squires, 1982).*

Animals	Location	Diet summary	Authors
Red kangaroos	North-west N.S.W.	Grasses (mainly *Eragrostis* spp. and *Enneapogon avenaceus*) when available; at other times chenopods (mainly *Bassia diacantha*) but not large chenopod shrubs.	Bailey *et al.*, 1971
Cattle and red kangaroos	Alice Springs N.T.	Cattle ate small amounts of a large number of species of grass; kangaroos ate larger amounts of a smaller number of species of grass.	Low *et al.*, 1973; Low & Low, 1975
Sheep, eastern grey and red kangaroos	Cunnamulla Qld	Mainly grasses eaten by both species of kangaroos; forbs and browse by sheep.	Griffiths *et al.*, 1974
Sheep, goats, euros and red kangaroos	North-west N.S.W.	Sheep ate mainly chenopods; goats ate trees and shrubs; euros ate grasses and shrubs; red kangaroos ate mainly grasses.	Dawson *et al.*, 1975
Sheep and goats	Ivanhoe N.S.W.	Sheep ate *Stipa variabilis* (grass) and *Bassia*; goats ate largely browse but on occasion some *Bassia*.	Wilson *et al.*, 1975
Cattle and sheep	Deniliquin N.S.W.	Sheep ate mainly annual medics and cool-season grasses; cattle ate more dry herbage and warm-season grasses.	Wilson, 1976
Red kangaroos, euros and sheep	North-west N.S.W.	All ate grasses and forbs in good seasons; selectivity increased as pasture dried out; reds selected grasses and chenopods; sheep selected flat-leaved chenopods (mainly *Atriplex* spp.); euros were highly selective of grasses.	Ellis *et al.*, 1977
Sheep	Julia Ck. Qld	Mitchell grass (*Astrebla*) was important throughout the year. Other grasses and forbs added to the diet.	Lorimer, 1978
Euros, goats, wallabies and rabbits	North-west N.S.W.	Euros ate grass; goats ate grasses and browse; wallabies ate forbs and browse; rabbits ate grass.	Dawson & Ellis, 1979
Sheep, cattle and goats	Coolabah N.S.W.	Sheep and cattle ate mainly grass; goats ate mainly shrubs although the diets overlapped on common plants.	Harrington, 1978

Table 5.3. *Diet studies on mammalian herbivores in the arid regions of Australia (after Squires, 1982).*

Animals	Location	Diet summary	Authors
Sheep, cattle and goats	Coolabah N.S.W.	Sheep ate mainly forbs and grasses; cattle and goats had a larger amount of browse in the diet.	Squires, 1980, 1981
Sheep and cattle	Booligal N.S.W.	Sheep ate mainly grasses; cattle ate chenopods.	Graetz & Wilson, 1980

Fig. 5.6. Diet of sheep and three species of kangaroos in good and dry times. The data are taken from studies in western New South Wales and southern Queensland (see text for references). G – grasses, F – forbs, C – chenopod sub-shrubs, B – chenopod shrubs, S – shrubs. Graphs on the left are for times of plenty, those on the right for dry times.

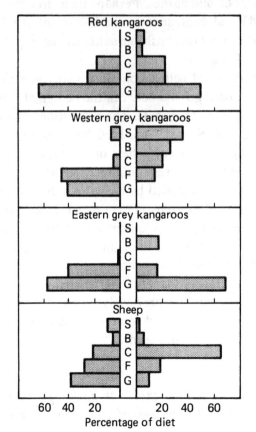

northwestern New South Wales and southern Queensland. The data used are means of the work of Griffiths & Barker (1966), Bailey, Martensz & Barker (1971), Griffiths *et al.* (1974), Dawson *et al.* (1975), Ellis *et al.* (1977) and this study. The figure shows that kangaroos and sheep eat a similar amount of grass and forbs in times of plenty but when conditions deteriorate the sheep more readily shift their diet to chenopod sub-shrubs, red kangaroos tend to keep the same relative proportions in their diet, eastern grey kangaroos select grasses and chenopod shrubs as a substitute for forbs and western grey kangaroos shift to eating shrubs (including chenopod shrubs).

Studies on red kangaroos show that they prefer to eat various species of grasses and forbs (Chippendale, 1962; Griffiths & Barker, 1966; Chippendale, 1968b; Storr, 1968; Bailey *et al.*, 1971; Low *et al.*, 1973; Griffiths *et al.*, 1974; Dawson *et al.*, 1975; Ellis *et al.*, 1977) and this was found also at Kinchega during times of plenty. As the vegetation dried off they switched to shrubs and bluebush although there was always some chenopod in the diet. Bailey *et al.* (1971) found chenopod sub-shrubs (mainly *Sclerolaena diacantha*) in the diet of reds at the height of a drought but in contrast to diets on Kinchega there were no perennial saltbushes or bluebushes. Perhaps their drought was not as severe as that of 1982-83 and consequently the bottom of the food selection table was not reached. Perennial bluebush and saltbush may represent the last dietary resort for the red kangaroo.

During this study I noticed that faecal pellets contained many seeds. At the peak of the drought the almost bare ground between the perennial shrubs was littered by pellets of kangaroo dung (Fig. 5.3). The vegetation regrowth in May 1983 can be seen in Fig. 5.4. Some of the seed for later regrowth must have been stored in the dung. The faecal pellets of kangaroos form a tough external coat as they dry and they can remain intact for many months unless that coat is broken by trampling or eroded by heavy rain. Seeds inside the dung are safe from ants. Briese & Macauley (1981) noted in their study of ant populations on the Riverine Plain that ants harvested only freshly fallen seed, and Briese (pers. comm.) commented that the mandibles of ants would be too weak to break the external coating of the pellets.

The passive strategy of using the faeces of native herbivores as a secure seed storage facility is a plant adaptation not previously noted in Australia. Sheep faeces are probably less important in seed storage because the sheep grind their food finer and thereby destroy more seeds.

5.4   Conclusions

A study of western grey and red kangaroos on Kinchega National Park in the years 1980-84 has shown that in periods of high biomass (usually following rain) there is little difference in the diets of the two species, both of which concentrate on eating various species of grasses and forbs. However when a drought occurs the plants eaten differ, the reds concentrating on chenopod shrubs such as black bluebush (*Maireana pyramidata*) and the greys on other shrubs such as prickly wattle (*Acacia victoriae*) and *Eremophila* spp.. These diet trends do not differ substantially from those obtained by other studies for sheep and kangaroos in the sheep rangelands of eastern Australia.

Neither species of kangaroos grind their food as finely as do sheep and this allows their faecal pellets to act as a seed store in drought periods.

6

Factors affecting food intake of rangelands herbivores

JEFF SHORT

6.1   Introduction

The amount of food eaten by a population of large mammalian herbivores affects their survival, reproductive success and, in the case of domestic stock, their economic performance. But intake by the herbivores also represents offtake from the pasture. Herbivores, by harvesting plant material from the pasture, have a feedback effect on biomass, growth and species composition of the pasture. Hence the herbivore and its rangelands pasture form an interactive relationship.

This chapter summarises the available information linking food intake of the major mammalian herbivores of the Australian sheep rangelands with food availability and identifies the factors, other than food availability, that modify intake. It then discusses the major feedback effects of this offtake on the vegetation.

The relationship between the food intake of a herbivore and vegetation biomass is known as the functional response. It can take a number of theoretical forms (Noy-Meir, 1975) but is most commonly expressed as an asymptoting function, often as a monotonically increasing curve (e.g. Allden & Whittaker, 1970; Arnold, 1975). At low biomasses the food intake of the herbivore is depressed because of the difficulties in locating and harvesting food. At high biomasses the intake of the herbivore is satiated and hence intake is relatively constant over a wide range of biomasses.

A knowledge of the functional response of a herbivore may indicate whether its rate of increase is constrained by food shortages. This is likely if food availabilities typical in the field fall below that necessary for the herbivore to satiate its hunger (i.e. the asymptote of the functional response curve). Functional responses obtained for several species feeding on the same pasture indicate the pasture biomass at which competition may limit each species' productivity. Competition occurs at pasture biomasses below that at which satiation occurs. Functional responses may be incorporated, also, into theoretical models of grazing systems in an effort to predict their dynamic behaviour after perturbation. For example, Caughley (1976a) used a model that included a functional response, to demonstrate that herbivores colonising a new habitat in a constant environment are likely to increase in number beyond that sustainable by their food supply (i.e. the ungulate eruption). In addition, the impact of herbivores on the pasture can be assessed by examining the ungrazable residue of vegetation and the steepness and the shape of the functional response curve (Noy-Meir, 1975).

The introduced sheep and rabbit and the native kangaroos are the major mammalian herbivores of the rangelands of southern Australia. All of the work on these species described below has been conducted in chenopod shrublands. These communities cover some 500,000 km² of southern Australia (6% of the land surface of Australia), and are dominated by species from the family Chenopodiaceae (Wilson & Graetz, 1979). Species are xeromorphic and halophytic with semi-succulent, reflectant and usually hairy leaves (Graetz, 1975). The shrubs are generally ⟨1.5 m in height and occur in extensive, uniform stands of one or two species on flat or very gently rolling landscapes (Wilson & Graetz, 1979). The major genera of economic importance to the pastoral industry include *Atriplex* (saltbushes), *Maireana* (bluebushes) and *Sclerolaena* (copperburrs).

6.2   The grazing trials

This section introduces a number of grazing trials that have been conducted to determine the food intake of either sheep, rabbits or kangaroos. It briefly describes methodology for each trial and the vegetation association in which the trials were conducted. Generalisations emerging from these trials are listed in a following section.

6.21   *Kangaroos*

Short (1985, 1986) has produced functional responses for red kangaroos grazing a pasture of annual forbs and grasses with scattered bluebush (*Maireana pyramidata*) and for red and western grey kangaroos grazing a pasture of annual grasses and forbs. Both trials were conducted in Kinchega National Park. The first was conducted on a pasture of senescing annuals with an initial biomass of 500 kg/ha. An additional 100 kg/ha of bluebush was present. The second trial was conducted on a pasture with an initial biomass of 1,000-1,200 kg/ha that was dominated by annuals. The biomasses present at the commencement of the grazing trials represented average and above average grazing conditions (Fig. 4.3). Animals progressively defoliated yards of 0.04 ha over a period of 12-18 days. Food intake was estimated as the difference between successive daily estimates of vegetation biomass corrected for trampling. Inverted exponential curves (Noy-Meir, 1979) were fitted by least-squares analysis to the data for each species. Results are reproduced in Fig. 6.1. Curves for red and western grey kangaroos differed substantially in the amount of vegetation left uneaten. No other information is available on the relationship between vegetation biomass and food intake for kangaroos.

6.22   *Sheep*

A number of studies have measured the food intake of sheep grazing pasture and shrubs within chenopod shrublands. Leigh, Wilson & Mulham (1968) and Wilson, Leigh & Mulham (1969) measured the food intake of merino sheep grazing shrublands of cotton-bush (*Maireana aphylla*) with a grass (*Stipa variabilis, Danthonia caespitosa*) understorey and a bladder

Fig. 6.1. The functional responses of red kangaroos feeding on: (a) a pasture of annual forbs and grasses and scattered bluebush *Maireana pyramidata*, $y = 62\{1\text{-exp}(-V/84)\}$; (b) a pasture of annual forbs and grasses, $y = 66\{1\text{-exp}(-V/34)\}$; and (c) the functional response of western grey kangaroos feeding on a pasture of annual forbs and grasses, $y = 87\{1\text{-exp-}[V\text{-}180)/270]\}$.

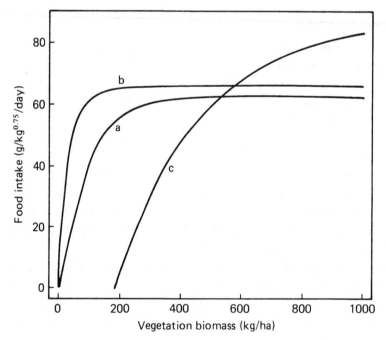

saltbush (*Atriplex vesicaria*) and cotton bush community respectively. In both studies sheep were set-stocked at a range of grazing pressures for several years. Food intakes were calculated seasonally over a 12-month period from measurements of *in vitro* digestibility and daily faecal output. Leigh & Mulham (1966a, b), Noble (1975) and Short (1985) conducted graze-down trials with merino sheep in chenopod shrubland. These were trials of short duration where sheep were placed in yards at very high densities so that their food intakes could be assessed at a wide range of plant biomasses ranging from the starting biomass to an ungrazable residue. The trials of Leigh & Mulham (1966a, b) were conducted in a bladder saltbush – cotton bush community and a cotton bush – grass (*Stipa variabilis, Danthonia caespitosa*) community in south-western New South Wales. Measurements of pasture biomass were made daily for 4-7 days in yards of 20 x 20 m (0.04 ha) to record the progressive defoliation by six merino wethers. Noble (1975) measured food intake of merino sheep feeding in a bladder saltbush – pearl

Fig. 6.2. The functional responses of sheep grazing chenopod shrubland: (a) Wilson, Leigh & Mulham (1969), *Atriplex vesicaria – Maireana aphylla* community, $y = 77 \{1\text{-exp}(-V/100)\}$; (b) Leigh & Mulham (1966a), *Atriplex vesicaria – Maireana aphylla* community plus *Maireana aphylla – Stipa variabilis – Danthonia caespitosa* community, $y = 61 \{1\text{-exp}-[-(V\text{-}150/325)]\}$; (c) Leigh, Wilson & Mulham (1968), *Maireana aphylla – Stipa variabilis – Danthonia caespitosa* community, $y = 59$; (d) Short (1985), *Maireana pyramidata* community, $y = 61.1 \{1\text{-exp}(-V/111)\}$; and (e) Noble (1975), *Atriplex vesicaria – Maireana sedifolia* community, $y = 83.4 \{1\text{-exp}(-V/426)\}$.

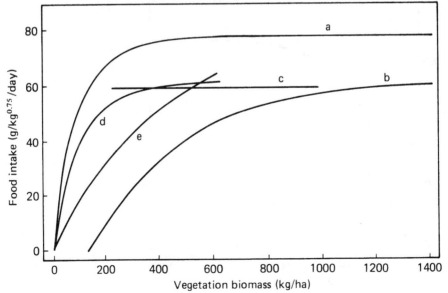

bluebush (*Maireana sedifolia*) community in South Australia. Six ewes defoliated an area of 0.1 ha in 6-7 days. Short (1985) produced a functional response for sheep grazing a bluebush (*M. pyramidata*) shrubland at Kinchega in western New South Wales. Two pairs of sheep progressively defoliated areas of 0.04 ha of bluebush and annual forbs and grasses over 12 days.

The data from all the above studies are presented in a common format (inverted exponential curve; [Noy-Meir, 1979]) in Fig. 6.2 to allow comparison. Vegetation losses in the graze-down studies were converted to food intake by correction for loss due to trampling. Curves were fitted by least squares analysis to the data of Leigh & Mulham (1966a, b), Wilson *et al.* (1969) and Noble (1975). All data points of Leigh *et al.* (1968) appeared to estimate the asymptote and hence a straight line through the mean parallel to the x-axis was fitted to these results. The data of Leigh & Mulham (1966a, b) and Noble (1975) were corrected by the latter's estimate of trampling and converted to $g/kg^{0.75}/day$.

6.23   *Rabbits*

Short (1985) has produced a functional response for rabbits grazing an annual forb and grass pasture interspersed with bluebush (*M. pyramidata*). This curve is reproduced in Fig. 6.3. This study was conducted at Kinchega

Fig. 6.3. The functional response of rabbits feeding on a pasture of annual forbs and grasses and scattered bluebush *Maireana pyramidata*, $y = 68 \{1-\exp(-V/138)\}$.

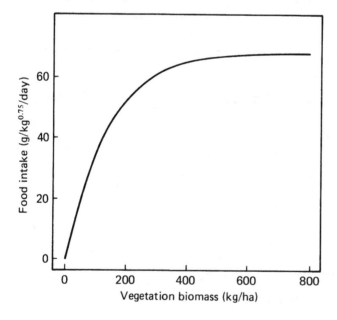

in conjunction with those for red kangaroos and sheep discussed above. Eight rabbits were held in each of two yards (10 x 10 m) with shelter and water provided.

6.3 Maximum food intake

Kangaroos, rabbits and sheep appear to eat approximately the same amount (60-80 g/kg$^{0.75}$/day) when intake is expressed on a metabolic weight basis and when food is not limited in availability. Absolute intake varies widely according to body size ranging from about 80 g/day for a rabbit (6-7% of body weight) to 1,200-1,500 g/day (2-3% of body weight) for merino sheep with kangaroos being intermediate between the two (700-1,000 g/day, 2-3% of body weight). Absolute food intakes of the three species are in the approximate ratio of 1:8.5:12 (rabbit : red kangaroo : sheep) assuming adult body weights of 1.5, 30 and 50 kg respectively.

A comparison of Figs. 6.1, 6.2 and 6.3 indicate that competition for food between red kangaroos, sheep and rabbits is minimal at biomasses above 300 kg/ha. Each obtains its *ab libitum* requirement at biomasses greater than this.

6.4 Grazing efficiency and vegetational residue after grazing

A high grazing efficiency is indicated by either a steep functional response or a small residue of vegetation. Efficient grazers can harvest a greater amount of food from an area at a lower biomass of vegetation. While no testable differences in grazing efficiency or vegetational residue could be measured between rabbits, red kangaroos and sheep at Kinchega it appears that there is a major difference between these species and western grey kangaroos, at least for the habitat in which the trials were conducted. Western greys appeared to be unable to reduce the pasture beyond 180 kg/ha whereas red kangaroos, sheep and rabbits were able to reduce pasture biomass to 20-50 kg/ha, depending on the biomass of unpalatable species present in the pasture (Fig. 6.1). Further experimentation and observation are required to discern whether this difference is real or whether it is an artefact of the grazing trial. The trial that compared western grey and red kangaroos was conducted in enclosures on red sandy soil, one of the two major soil types on Kinchega. Western greys utilise this habitat but apparently prefer the more densely timbered associations of the grey soil on the Darling River floodplain (Chapter 7). Both red and grey kangaroos concentrate on grey soils when pasture is scarce (Chapter 7), hence their grazing efficiencies on this soil type may be more critical to their survival in drought when pasture is scarce.

There may also be differences between sheep and cattle. Wilson (1974) reported that graziers on the north-west Riverine plain of New South Wales considered that cattle must be removed earlier than sheep at the onset of drought because of their lesser foraging abilities. When pasture is sparse, grazing cattle experience great difficulties in food prehension (e.g. Graetz & Wilson, 1980). At low biomasses the deficiency in herbage yield cannot be compensated by increasing the area of pasture available to the animals (Allden & Whittaker, 1970).

6.5   The effect of food availability

Intake varies with the availability of food. The relationship between food intake and vegetation biomass is best represented by an asymptoting function. At moderate to high biomasses food intake is relatively constant but at low biomasses food intake is depressed. The functional response of herbivores grazing rangelands has a similar form to those grazing temperate sown grasslands (Van der Kley, 1956; Willoughby, 1959; Arnold & Dudzinski, 1967a, b; Allden & Whittaker, 1970). The herbivore's food intake declines with declining plant biomass because it is forced to take bites of smaller size. For example, the bite size of cattle grazing a tropical pasture was reduced from 300 to 75 mg organic matter per bite due to declining food availability. Small increases in grazing time (550 to 650 minutes per day) and biting rate (55 to 62.5 bites per minute) were insufficient to compensate and food intake declined (Chacon & Stobbs, 1976). Sheep and kangaroos showed a greater ability to increase grazing time (e.g. sheep: 6.7 to 12.5 h/day, Allden & Whittaker, 1970; red kangaroos: 7.5 to 16 h/day, Short, 1986) but this may still be insufficient to maintain intake as vegetation biomass falls.

6.6   The effect of food quality

Low quality food depresses intake in ruminants and macropods by slowing the rate of passage of digesta through the gut (Foot & Romberg, 1965; McIntosh, 1966). There is a significant linear relationship between food intake and food quality (as measured by nitrogen content as a percentage of dry weight) of red kangaroos and sheep fed foods of different quality in metabolism cages (Fig. 6.4). An increase in the amount of nitrogen in the diet from 1 to 2% of dry weight produces an increase in food consumption of 35-40% in both species. More than half of the difference in asymptotes of food intake in sheep between the studies of Leigh *et al.* (1968) and Wilson *et al.* (1969) (60 *v.* 77 g/kg$^{0.75}$/day, Fig. 6.2) can be attributed to differences in food quality. Food material collected by oesophageal fistulas from the sheep of Wilson *et al.* (1969) were significantly higher in nitrogen content (2.3 *v.* 1.7%; $F_{1,12} = 9.57$, $P\langle 0.01)$ and *in vitro* digestibility (60.9 *v.* 50.6%; $F_{1,12} = 4.57$, $0.05\langle P\langle 0.1)$.

The density of herbivores that a pasture can support may be reduced by an increased representation in the pasture of species of low quality and palatability. In the arid north-west of Western Australia a combination of high stocking levels of sheep through a series of droughts culminating in that of 1935-6, and the failure to spell the pasture after drought-breaking rains caused a major shift in species composition of the pasture (Suijdendorp, 1955; Ealey, 1967a). Spinifex *Triodia pungens*, which is low in palatability and in nitrogen content (Storr, 1968), replaced many of the more nutritious grasses and forbs. Spinifex provides forage of a quality sufficient for maintenance of sheep but not for lactation. This led to a decline of approximately 50% in the numbers of sheep within the district, largely through reduced reproductive success (Ealey, 1967a).

An important aspect of quality of diet for the rabbit is the availability of water in food. Rabbits are less mobile than the larger herbivores and drinking water is often absent within their range, particularly in summer. Hence they must obtain all their water requirements from food. Cooke (1982) found that rabbits were not able to reduce their requirements for water to less than 55% of their total intake of food and water. Hence when drinking water is absent and when pasture species frequently contain only 10-15% water (i.e. in summer or in drought), rabbits depend on the presence of succulent perennial vegetation.

Fig. 6.4. The effect of food quality (% nitrogen in the diet) on food intake of red kangaroos and sheep. Data are from studies by Foot & Romberg (1965); McIntosh (1966) and Forbes & Tribe (1970).

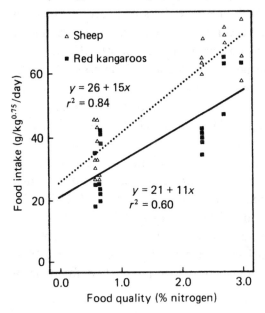

Rabbits, by virtue of being smaller than sheep and kangaroos, may be less able to cope with a decline in food quality. A herbivore's ability to obtain energy from a low quality (high fibre) diet is dependent largely on the volume of its fermentation chambers. The volume available is directly related to body weight, yet the herbivore's energy requirement is related to weight to the exponent 0.75 (Degabriele & Dawson, 1979). As herbivore size decreases energy requirements per unit weight increase but fermentation contents per unit weight remains constant. For example, the fermentation contents of the gut of the rabbit are only about 75 $g/kg^{0.75}$ of body weight as compared to 368 $g/kg^{0.75}$ for sheep (Parra, 1978). Rabbits are only 60% as efficient as ruminants in digesting fibre (Hintz, 1969) and must feed on a high quality diet ($\langle 40\%$ fibre; Cooke, 1974) to obtain their energy requirements.

6.7   Relative preferences for chenopod shrubs and pasture

The rate of consumption of chenopod shrubs is less than expected by their abundance as forage within the pasture. Leigh *et al.* (1968) found that *Maireana aphylla* contributed little to the diet of sheep (a maximum of 2%) despite it contributing more than 20% to the available forage. Similarly, when forage was abundant sheep selected less than 17% of their intake as *Atriplex vesicaria* despite it being the major constituent (65-70%) of the pasture (Wilson *et al.*, 1969). The chenopods *Maireana sedifolia*, *M. pyramidata* and *M. sclerolaenoides* were the least preferred species in the trial of Noble (1975) but some *Sclerolaena* species (*S. biflora*, *S. uniflora* and to a lesser extent *S. obliquicuspis* and *S. patenticuspis*) and *Atriplex vesicaria* were preferred to mature grasses of *Stipa nitida* and *Danthonia caespitosa*. The percentage of *Maireana pyramidata* in the diets of sheep, rabbits and red kangaroos were less than their percentage occurrence in the pasture at the commencement of a graze-down trial conducted by Short (1985).

However in all trials herbivores responded to declining biomasses of vegetation by broadening their range of acceptance to include the chenopod shrubs. Often they defoliated them totally. Hence chenopod shrubs are valuable as a reserve supply of food during drought. They may become the principal or even the sole dietary item of sheep in drought (Wilson & Graetz, 1979). Sheep incorporate a greater amount of chenopod in the diet at higher biomasses of vegetation than do either rabbits or kangaroos. Sheep consumed 20% of their intake as *M. pyramidata* at a pasture biomass of 250 kg/ha compared to only 10% by rabbits and red kangaroos at this biomass (Short, 1985). The lesser ability of kangaroos to consume chenopod shrubs was reflected in their consumption relative to that of sheep during the 1982-3 drought on Kinchega and Tandou. Kangaroos proved unable to defoliate large areas of *M. pyramidata* shrubland at Kinchega National Park. Fifty percent of kangaroos died (Chapter 8) yet there was no noticeable impact on the shrubland. This contrasted with major defoliation of bluebush by sheep on the adjoining properties.

Chenopod shrubs vary in their palatability between species. Sheep held in metabolism cages fed a diet of dried leaves of a single species of chenopod favoured *Sclerolaena diacantha* (1086 g of organic matter (OM) per day) over *Atriplex vesicaria* (671 g OM/day), *A. nummularia* (432 g OM/day) and *Maireana pyramidata* (392 g OM/day) (Wilson, 1977). The intake for *Maireana pyramidata*, the dominant chenopod shrub on Kinchega National Park and the adjoining sheep station Tandou, represents only 36% of the intake of sheep fed lucerne hay and confined in a similar manner (Foot & Romberg, 1965; McIntosh, 1966). Fluctuations in sodium and chloride concentrations within a species may affect the acceptability of foliage of that species to herbivores. Sodium and chloride concentrations in the leaves of *A. vesicaria* and *A. nummularia* in late spring, summer and early autumn reached 20-25% of dry weight, twice those of the remainder of the year (Sharma, Tunny & Tongway, 1972). Hence chenopod shrubs are likely to be least palatable in summer when food is often in short supply.

6.8   Other factors affecting food intake

A number of other factors have been shown to affect the food intake of herbivores in other environments and may play a role in the rangelands. These include:

6.81   *Structure of the vegetation*

The structure of vegetation may either aid or inhibit the ability of a herbivore to feed. Black & Kenney (1984) found that, at herbage availabilities less than 1,000 kg/ha, the intake rates of sheep were several-fold greater when grazing tall sparse pastures as compared to short, dense pastures. This explains the large differences between functional response curves obtained for sheep in rangelands, where pastures are tall and sparse (Fig. 6.2), and temperate sown pastures where swards are dense (e.g. Willoughby, 1959; Arnold & Dudzinski, 1967a, b; Allden & Whittaker, 1970). In the former, food intake declines at biomasses below 600 kg/ha whereas in the latter the decline is at biomasses over 1,000 kg/ha.

Sheep grazing a pasture of *Phalaris tuberosa*, annual grasses and subterranean clover had a depressed intake at high pasture biomasses ()1,000 kg/ha) as the large amount of dry material in the pasture hindered them from harvesting green material (Arnold, 1964). The green *Phalaris*, which would normally be preferred, was enmeshed in dry grass so that almost none was eaten. The lower intake was due largely to a reduction in biting rate from 60-80 bites per minute on a short green pasture to a rate of 25-40 bites per minute on abundant dry pasture.

Allden & Whittaker (1970) altered the close relationship between pasture yield per unit area and plant height by ploughing a Wimmera ryegrass (*Lolium rigidum*) pasture to leave strips of 25 cm width. Rate of intake was closely related to tiller length, falling away at lengths less than 20 cm. The relationship did not change with treatment (no cultivation; bare areas of 45 cm width; bare areas of 90 cm width) whereas that of intake versus dry matter yield (kg/ha) did. The authors concluded that length of tiller is much more closely related to rate of intake than is dry matter per unit area. Arnold (1975) found that food intake of sheep grazing a pasture of *Phalaris tuberosa* – *Trifolium subterraneum* was depressed at leaf lengths below 10 cm. Intake fell to 50% of maximum at a leaf length of 2.5 cm. A declining proportion of leaf to stem in a tropical pasture of legumes may greatly reduce food intake in dairy cattle because of their unwillingness to feed on stem (Chacon & Stobbs, 1976; Hendricksen & Minson, 1980).

Red kangaroos in central Australia favour short green grass (Chippendale, 1968b) over long, dry grass. When kangaroos feed in long, dry grass they eat small green shoots sprouting from the bases of green clumps or green herbage growing between the clumps (Newsome, 1971). This tendency to feed at the bottom level of the available pasture may well slow their rate of intake. Newsome claimed that cattle facilitated grazing by kangaroos by consuming the long, coarse material forcing it to sprout green shoots. This is similar to the facilitation observed between African ungulates in the Serengeti by Gwynne & Bell (1968).

6.82   Lactation

The burden imposed on the female by advanced lactation may substantially increase food intake. Prince (1976) found that the nutritional burden imposed by an advanced pouch young on a female kangaroo could easily exceed 0.5 times the apparent minimum maintenance level for that female. Similarly, lactating Border Leicester and Merino ewes have a digestible organic matter intake 40-45% greater than that of comparable dry ewes (Arnold & Dudzinski, 1967a).

6.83   Shearing

Shearing of sheep in late spring increased food intake (digestible organic matter intake) by 42-62% (Wheeler, Reardon & Lambourne, 1963).

Sheep ate 300-400 g/day more after winter shearing than unshorn sheep at all levels of pasture biomass (Arnold, 1964).

6.9 Discussion

To this point I have discussed the food intake of kangaroos, sheep and rabbits as a function of food availability (the functional response) and the effect of factors other than vegetation biomass on food intake. In the remainder of this chapter I will discuss several broad ecological questions upon which the above relationship impinges.

(1) Are the rates of increase of herbivores on Kinchega and Tandou constrained by the availability of food?

This question can be addressed by comparing vegetation biomasses typical in the field to the herbivore's functional response. If the biomass of vegetation falls below that at which the herbivore can harvest its *ad libitum* requirements then it is likely that the species' rate of increase is constrained by availability of food. The food intakes of rabbits, red kangaroos and sheep in the Kinchega trial began to drop off at a biomass of 250-300 kg/ha. This corresponded to a pasture biomass (excluding browse) of about 180 kg/ha. Pasture biomass was below 180 kg/ha for 47% of the three years of the Kinchega study (Fig. 6.5) including the 15-month period between February 1982 and May 1983

Fig. 6.5. The estimated relative food intake of red kangaroos and the biomass of pasture at Kinchega National Park, 1980-4. Pasture biomass estimates are from Chapter 4. Food intakes are calculated using the equation $y = 66 \{1-\exp(-V/34)\}$. Forty percent of red kangaroos on the Park died between November, 1982 and February, 1983 (Chapter 8).

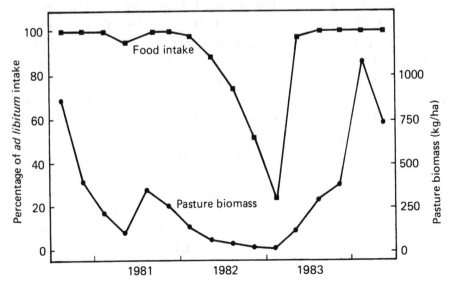

(Chapter 4). Condition indices of red kangaroos (Chapter 9) were low over this period due to their reduced food intake. Food intake of red kangaroos during this period fell to 50% of *ad libitum* in November 1982 and 25% in February 1983 (Fig. 6.5). The food intake of rabbits and the unsupplemented food intakes of sheep would have fallen to similar values. Hence, it is likely that the rates of increase of all three species are constrained by shortages of food during troughs in production of vegetation, with rates of increase becoming negative during prolonged dry periods when food becomes increasingly scarce. This in fact occurred. Approximately 40% of red kangaroos died over the summer (Chapter 8), rabbit numbers fell by 84% between May 1982 and February 1983 on Kinchega (G. Robertson & H. Armstrong, unpublished data) and sheep, which were hand fed from May to August 1982, were transferred from the study area on Tandou to the crop stubble on the dry bed of Lake Tandou. While periodic food shortages constrain the rate of increase of the herbivores, other agents of mortality also operate (e.g. predation in rabbits; Wood, 1980).

(2) What aspects of the feeding behaviour or the functional responses of the herbivores promote or inhibit loss of plant or animal species or major changes in the ratios of plant groups (annual grasses and forbs, perennial grasses, chenopod shrubs, trees) or dampen or accelerate fluctuations in biomass of either plants or animals?

Red kangaroos, sheep and rabbits have a higher grazing efficiency on rangelands pasture compared to that of sheep (and probably kangaroos and rabbits) grazing temperate sown pastures. They can maintain their intake to a lower pasture biomass because of the structure of the vegetation of rangelands, it being taller for a given biomass per hectare than temperate pasture (Leigh *et al.*, 1968). Noy-Meir (1975) and Caughley (1976a) suggested that an increase in grazing efficiency leads to a decrease in stability of the plant-herbivore system (i.e. greater fluctuations in both biomass of vegetation and herbivore numbers around an equilibrium with the possibility of herbivores going to extinction). The high grazing efficiency of sheep on rangelands is reflected in the suggestion by range scientists that decisions to reduce stocking density should be based on the condition of the vegetation rather than the condition of the animals (Stoddart, Smith & Box, 1975 p. 271; Graetz & Wilson, 1980).

Red kangaroos have both a steep functional response and a propensity to dig for underground stems when food is in short supply. In the grazing trial at Kinchega they dug vigorously for below-ground stem at biomasses of 200 kg/ha and below. Most of the biomass remaining at this time consisted of *Sclerolaena diacantha* and *Babbagia acroptera*. Both species are chenopods with high salt contents and relatively low palatability. This ability to harvest below-ground vegetation may prejudice the survival of palatable perennial pasture species by eliminating their potential sources of regeneration.

Grazing at high densities by domestic stock is regarded as the dominant process in reducing or eliminating summer-growing perennial grasses in higher rainfall areas of the sheep rangelands in Australia (Moore, 1962). Grazing by sheep at high densities has almost eliminated kangaroo grass (*Themeda australis*), native oat grass (*T. avenacea*) and plains grass (*Stipa aristiglumis*) from many of the woodland communities (e.g. rosewood, *Heterodendrum oleifolium* — belah, *Casuarina cristata*) to the east and south of Kinchega. These are tall, summer-growing perennials. They may be severely defoliated in dry summers when food for sheep is often scarce and are dormant in winter when food is commonly available (Williams, 1968). They are replaced by annuals such as introduced burr medics and barley grass *Hordeum leporinum* in winter and burr grass (*Tragus australianus*) and button grass (*Dactyloctenium radulans*) in summer (Moore, 1962). Westoby (1979-80) suggests that set-stocking in rangelands disadvantages perennials relative to annuals. Annuals complete their life cycle after rain when food is abundant and grazing pressure moderate but perennials, which continue to photosynthesise long into drought, may be the major food source when annuals are represented only by seed. Hence the intensity of grazing on the perennial component may be much greater than that on the annual component. This effect will be enhanced by grazers capable of maintaining their intake at very low biomasses, whether they be domestic stock or native fauna. It may be enhanced also by graziers maintaining stock numbers during drought by supplementing their intake by hand-feeding and lopping of shrubs and trees. However, note that there are numerous rangelands where annuals continue to dominate in the absence of grazing (Westoby, 1979-80) and that severe mortality of perennial grasses and major changes in the perennial dominant may take place due to climatic variation in the absence of grazing (Williams & Roe, 1975).

Annuals survive as seed during drought and escape the high grazing pressures exerted on the perennials within pastures of declining biomasses. However, sheep supplement their food intake by eating the seeds of annual medics when vegetation is scarce (Leigh *et al.*, 1968). No information is available on the effect of this on subsequent biomasses of these species after rain. Grazing may also reduce the seedcrop of annual plants. *Enneapogon avenaceus*, a short-lived annual or biennial grass common in western New South Wales, requires that 60% of plants in the population be ungrazed or only lightly defoliated to produce sufficient seed for continued persistence (Bosch & Dudzinski, 1984). However in this study no account was taken of the role of the herbivore (in this case, cattle) in seed dispersal. Barker (Chapter 5) presented evidence that kangaroos may pass a higher proportion of seeds intact compared to sheep and hence may be more effective seed dispersers.

Concern has been expressed regarding the impact of herbivores (particularly sheep) on the chenopod shrublands (Newman & Condon, 1969; Wilson & Graetz, 1979). Sheep eliminated 98% of *Atriplex vesicaria* bushes from the pasture in a two-year grazing trial on the Riverine plain of New South Wales at a stocking rate of 1.2 sheep per hectare, twice the district average (Wilson *et al.*, 1969). Lay (1979) reported a significant loss of *Maireana sedifolia* bushes over a 22-year period on properties in northern South Australia due to browsing by sheep. Most shrub loss occurred around watering points. Crisp (1978) also found that *A. vesicaria* and *M. sedifolia* declined in number after heavy grazing by sheep.

The high searching efficiencies of sheep, rabbits and kangaroos allow them to find and eat regenerating shrub and tree seedlings. Sheep and rabbit grazing prevents recruitment of mulga, *Acacia aneura* (Crisp, 1978) and *A. burkittii* (Crisp & Lange, 1976), in South Australia and has reduced the density of myall, *Acacia pendula*, in the Riverina of New South Wales (Moore, 1953 *b*). Sheep paddocks in the arid zone of South Australia with no history of overstocking show total suppression of seedlings of species such as *Acacia sowdenii* over the past 100 years of pastoralism despite major and minor recruitment events (Lange & Willcocks, 1980). Lange & Willcocks (1980) demonstrated that sheep at typical densities in chenopod shrubland were able to locate and eat 'simulated seedlings' (pellets of *Medicago sativa*) scattered throughout a 4,050 ha paddock. Pellets did not exceed 1:200,000 of fodder biomass on offer in the paddock. Sheep ate 80-100% of the pellets over more than 75% of the paddock in seven days. The authors inferred from this result that sheep have the capacity to eliminate any sparse perennial plant species by seedling suppression, providing they are as attractive to sheep as the pellets. Rabbits are thought also to reduce or prevent the regeneration of *A. aneura*, *A. kempeana* and *A. papyrocarpa* in arid areas of South and Central Australia (B.G. Lay & M.H. Friedal, pers. comm. to Wilson *et al.*, 1984). Kangaroos at high densities are thought to prevent the regeneration of shrubs (e.g. *Hakea vittata*) and herbs (e.g. *Psoralea tenax*) in Hattah-Kulkyne Park in north-western Victoria (Cheal, 1984). The prevention of regeneration of shrub and tree species may have a major impact on the appearance of the landscape as mature individuals age and die.

6.10   Conclusions

(1) The food intake of a herbivore when plotted against vegetative

biomass forms an asymptoting function known as the functional response. Red kangaroos, sheep and rabbits at Kinchega eat about 60-80 $g/kg^{0.75}/day$ in chenopod shrubland when food is not limiting. Food intake begins to decline at a biomass of about 250-300 kg/ha. This suggests that there is minimal competition between these species at biomasses above 300 kg/ha.

(2) Food intake is affected also by food quality, the structure of the vegetation, varying palatabilities of pasture plants and by the physiological state of the animal (i.e. lactation, cold stress associated with shearing). These factors may have a considerable effect on intake. A change in food quality from 1 to 2% nitrogen content, for example, may increase food intake of red kangaroos and sheep by 35-40%. Lactation may increase a female's requirements for food by up to 50%.

(3) A comparison of the pasture biomasses present at Kinchega and Tandou from 1980-83 with the functional response of sheep, kangaroos and rabbits, suggests that the rates of increase of these herbivores are constrained by the availability of food. Food shortages during the 1982-3 drought led to major mortality of kangaroos and rabbits. Sheep were removed from the study area to reduce mortality.

(4) The high grazing efficiency of sheep, red kangaroos and rabbits, the tendency of red kangaroos to dig for below-ground stem and of rabbits to ringbark shrubs when food is in short supply, and the predilection of sheep and rabbits for eating shrub and tree seedlings may be potent forces in changing the structure and species composition of the rangelands. Maintenance of sheep numbers during drought by management practices such as the lopping of shrubs and hand feeding may exacerbate their impact on the vegetation.

7

The mobility and habitat utilisation of kangaroos

DAVID PRIDDEL

7.1   Introduction

This chapter investigates the mobility, home range and habitat utilisation of red and western grey kangaroos on Kinchega National Park and Tandou between 1979 and 1982. Of particular interest was how far kangaroos moved within Kinchega, how this compared with movements outside Kinchega, and to what degree, if any, these movements were constrained by the fence surrounding Kinchega.

Red kangaroos inhabit the harsh dry environment of inland Australia. Australians think of them as highly nomadic creatures that regularly move hundreds of kilometres in search of greener pastures. This high mobility is consistently invoked to account for the marked changes that are sometimes observed in their dispersion.

Scientific evidence conflicts with this view. There is no doubt that some kangaroos move long distances. Bailey (1971) recorded that red kangaroos moved up to 216 km away from his study area in north-western New South Wales during a severe drought. Denny (1982) recorded that a red kangaroo moved from Tibooburra in New South Wales to Lake Frome in South Australia — a distance of more than 300 km. However, many kangaroos remain in the same area for several years. Frith (1964) sighted a conspicuous group of red kangaroos eight times from November 1960 to August 1961. All sightings were within a circle of 5 km diameter. Bailey (1971) tagged 143 kangaroos. Twenty-eight were sighted within 8 km of the release site after a period of six months. Denny (1982) tagged nearly 700 kangaroos caught at watering points. Fifty-eight percent of those sighted after release (no time period stated) were at their original place of capture.

From the pattern of movement of individual red kangaroos, Bailey (1971) concluded that there was a spectrum of mobility types in the kangaroo population studied, ranging from highly mobile animals to sedentary animals. Observations of the movements of red kangaroos reported in other studies appear consistent with this conclusion. Denny (1982) found wide-ranging kangaroos to be predominantly young males, suggesting that this particular sex and age class may be the means for long-range dispersal of the species. However, the proportion of the population confined to a home range and the proportion that is nomadic have not been determined; nor is it known whether these proportions vary as a function of weather. Although most evidence suggests otherwise, nomadism has tended to be accepted as the norm rather than the exception.

Studies of the eastern grey kangaroo, *M. giganteus*, in the eastern part of its range (Kirkpatrick, 1967; Jarman & Taylor, 1983) and of the euro, *M. robustus*, (Ealey, 1967b; Croft, 1980) have shown these species to be relatively sedentary, individuals rarely moving more than a few kilometres even during drought. Apart from the study reported in this Chapter, mobility of western grey kangaroos has not been studied.

Studies by Frith (1964), Newsome (1965a, b), and Bailey (1971) have shown that red kangaroos change their dispersion in response to the distribution of green pasture. Bailey (1971) noted that red kangaroos aggregated on green pastures when they were available, but that in dry conditions they favoured areas of long dry grass. He concluded that dispersion was primarily influenced by pasture quality. Johnson & Bayliss (1981) investigated the use of habitats by kangaroos on Kinchega and explained shift in habitat utilisation as a response to shifts in pasture productivity, with red kangaroos consistently seeking out areas of green pasture.

During drought, red kangaroos in Central Australia inhabited open plains or sheltered in woodland bordering those water courses and open plains where green pasture persisted (Newsome, 1965a). When green pasture became widespread, kangaroos moved from the open plains into wooded areas, apparently preferring the shelter of woodland when green herbage grew there.

7.2   Mobility of red and western grey kangaroos

It is important for both the control and conservation of kangaroos to know whether the individuals of a population are confined to a circumscribed area or home range, or whether they are nomadic. If kangaroos are nomadic it may be inappropriate to conserve them in relatively small parks and reserves. Similarly, measures initiated to control local kangaroo numbers can be thwarted if there are periodic influxes of kangaroos from considerable distances away.

A study of the mobility of red and western grey kangaroos was carried out in and around Kinchega National Park. The aims were twofold. First, to assess the mobility of individual kangaroos to determine whether they were sedentary or nomadic; and second, to determine whether mobility differed between species, between sexes, or between those kangaroos on Kinchega and those on Tandou.

Sedentary is used here to describe individuals confined to a circumscribed area or home range: nomadism implies the absence of a home range. 'A measure of the mobility of an individual does not itself indicate whether the individual is sedentary or nomadic; nomadic individuals do not necessarily range over larger areas than those restricted to a home range. Instead nomadism must be detected by the relationship between the extent of movement and time' (Caughley 1977, p. 62). For nomadic individuals, the distance from their place of capture increases with respect to time since capture.

Methods
Between July 1979 and November 1980, 261 red kangaroos and 170 western grey kangaroos were caught by 'stunning' (Robertson & Gepp, 1982). They

Fig. 7.1. A collar being put on a large red kangaroo.

were then tagged with individually recognisable collars, and released. Although the collars were pliant, they were not elastic and therefore could not be put on juveniles or small males.

A chronicle of the movements of each kangaroo was compiled from subsequent re-sightings. Several methods of obtaining sightings of tagged kangaroos were employed; none were systematic. Most sightings were the result of chance observations, but surveys to search for tagged kangaroos were conducted when time was available. A total of 112 volunteers walked a cumulative distance of more than 1,700 km along predetermined compass bearings to search for tagged kangaroos. Several collars were retrieved from kangaroos shot outside the study area. In all, 1,596 sightings were recorded between July 1979 and September 1982. From this information the distance moved by each kangaroo was calculated and the proportion of the kangaroo population that was sedentary and that which was nomadic was determined.

Results
Professional kangaroo shooters shot 11 tagged kangaroos (nine red females, one western grey female and one western grey male) outside the study area.

Fig. 7.2. Release of a male red kangaroo. Reflective symbols on the collar allow individual identification.

To date, the longest known movement of either species away from Kinchega is 285 km by a red female and 85 km by a western grey male. Most of the recorded movements away from the study area were in a north-easterly direction (Fig. 7.3). Whether this was the predominant direction of dispersal or whether it was the result of a sampling bias is unknown. It was the area most intensively worked by kangaroo shooters and therefore tagged kangaroos would have had a greater probability of being shot there than elsewhere.

Most tagged kangaroos were sighted infrequently and irregularly. Estimating the mobility of kangaroos from such casual sporadic sightings had numerous problems. A realistic estimate of the size of the home range was not possible and the mobility of an individual was estimated only in terms of the maximum distance between any two sightings. The number of sightings of each tagged kangaroo varied enormously, but the distance moved by an individual was not correlated with the number of sightings of that individual ($r_{166} = 0.10$ for reds and $r_{119} = 0.03$ for western greys).

Of red kangaroos tagged, 67% of females and 44% of males were sighted after release, and of these, 90% were never observed to move more than 9 km (Fig. 7.4). Of western grey kangaroos tagged, 71% of each sex were sighted after release and of these, 90% were never observed to move more than 6 km (Fig. 7.4). What happened to animals that were not sighted after release is unknown. They may have died, remained unseen in the study area, or moved away.

Fig. 7.3. Locations of marked kangaroos shot outside the study area. All red kangaroos were females. An additional red female was also shot near Lake Frome in South Australia 285 km to the north-west of Kinchega.

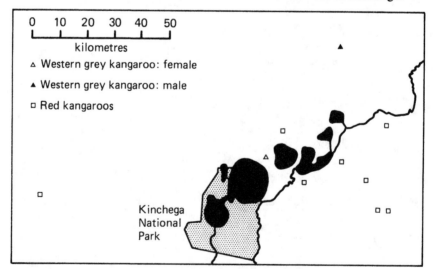

Although there was large variation in the mobility of individuals, red kangaroos were on average more mobile than western greys: the median of the distance moved was 6 km for reds and 3 km for western greys.

For western greys the frequency distribution of the distance moved differed little between sexes. A direct comparison between sexes of red kangaroos could be misleading; for reasons unknown only 44% of red males tagged were sighted again.

Mobility of kangaroos tagged on Tandou was compared with that of kangaroos tagged in an area of similar landform on Kinchega (Fig. 7.5). There was no difference in the frequency distribution of the distance moved by red kangaroos, but the frequency of western grey kangaroos that moved more than 4 km was less on Tandou. This may be an artefact of the small sample size on Tandou ($n = 14$).

Thirteen tagged kangaroos were known to have crossed the macropod-deterrent fence to get from Tandou into Kinchega. Four were known to have

Fig. 7.4. Distances moved by tagged kangaroos (n is the number of kangaroos tagged, n_0 is the number not sighted subsequent to tagging).

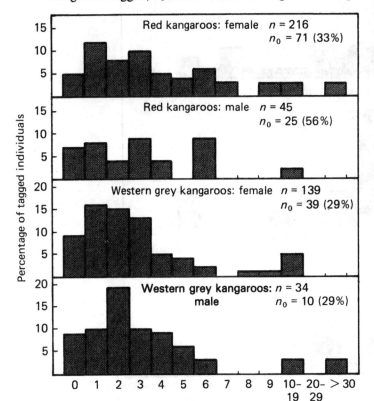

Maximum distance moved (km)

crossed in the opposite direction. Three kangaroos were recorded as making repeated crossings back-and-forth across the fence. One individual crossed on at least five occasions, and its home range must have included areas in both Kinchega and Tandou.

To examine whether tagged kangaroos were sedentary or nomadic, the distance between the most recent sighting and the capture location was plotted against the number of days since capture (Fig. 7.6). Movements of 20 km or more by nine red kangaroos were atypical of most individuals of the population.

Fig. 7.5. Frequency distributions of the distances moved by tagged kangaroos in Kinchega and Tandou (n is the number of kangaroos tagged, n_0 is the number not sighted subsequent to tagging).

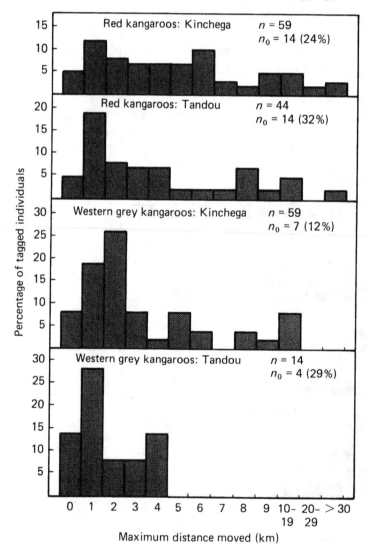

Data from these wide-ranging individuals, although plotted in Fig. 7.6, were omitted from subsequent analyses.

The distance moved by red kangaroos was regressed against the number of days since capture. In an attempt to ensure that the locations of capture and sighting were independent of one another, only those kangaroos that were sighted 100 days or more after capture were included. The slope of the regression was not significantly different from zero ($t_{130} = 1.137$), indicating that most individuals (132 of the 141 tagged) did not move progressively further from their place of capture and hence were not nomadic. Instead they were sedentary, being confined to home ranges. An identical analysis of Frith's (1964) data by Caughley (1977) also found red kangaroos not to be nomadic. The sedentary nature of most red kangaroos is highlighted by result from 22 individuals sighted 1,000 days or more after capture. Of these, 17 were within 7 km of their place of capture.

Western grey kangaroos were sedentary also (Fig. 7.6). The slope of the regression of distance moved by kangaroos sighted more than 100 days after capture did not differ significantly from zero ($t_{97} = 0.112$).

A few individuals of both species were more mobile (Figs. 7.3 and 7.6); 26 moved 10 km or more. These long movements were not confined to any discrete period of time. Furthermore, those individuals that were at times highly mobile, were not always so. For example, one red kangaroo was sighted 13 times over a period of 21 months and each sighting was within one kilometre of where it was caught; six months later it was shot 67 km away. Of the 26 highly mobile individuals 17 were reds and 9 were western greys. The ratio of females to males (corrected for the number of each sex tagged) was 3.5:1 for reds and 1:1 for western greys. These individuals all inhabited different areas and were caught in different locations. It seems that they responded to some particular social pressure, environmental stress, or localised disturbance that did not affect other individuals similarly. Whatever the cause, red females were most susceptible.

Clearly, it is inappropriate to generalise that kangaroos are either nomadic or sedentary. Although most are sedentary, being confined to a home range, a small proportion of the population is highly mobile at times. During the study on Kinchega, the proportion of highly mobile or wide-ranging kangaroos was less than 7%. Denny (1982) found this proportion to be as high as 20%, and suggested that it may increase further during dry periods.

7.3   Home ranges of red and western grey kangaroos

If kangaroos are confined to a home range then this home range must have limits and must therefore be measurable. By using radiotelemetry

Fig. 7.6. Distance kangaroos moved from their place of capture in relation to time since capture.

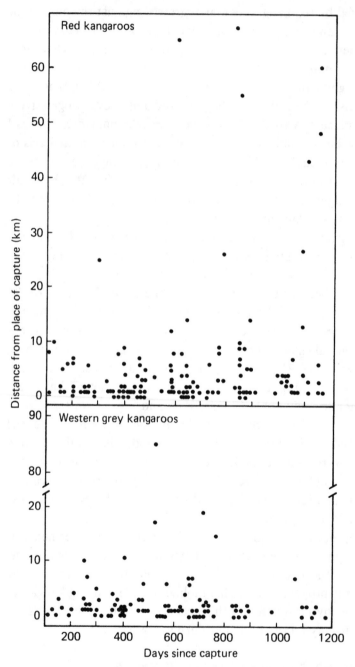

it was possible to estimate the size of the home range of 48 individuals. These estimates were then compared between species, between sexes, and between Kinchega and Tandou. They were also examined to determine whether home range changed in size over time, and whether any such changes coincided with changes in weather or pasture.

Methods

Forty-eight adult kangaroos were caught by 'stunning' (Robertson & Gepp, 1982). They were fitted with radio transmitters and released at the place of capture. They comprised six males and six females of each species on each of Kinchega and Tandou. Forty-two were caught in September 1980 and six were caught in December of the same year.

Transmitter life was approximately 20 months. During this time each individual's movements were traced from a network of four fixed tracking stations and an aircraft. The design, operating procedures, and accuracy of the radiotracking system are described in Priddel (1983).

Each kangaroo was radiotracked on four nights, each three or four days apart, for four seasons. Kangaroos were located hourly between 1900 h and 0700 h the following day. Previous continuous tracking showed that outside these hours kangaroos moved little. The individual's home range size was estimated by two techniques. The first was based on the minimum convex polygon (Southwood, 1966) and the second on the utilisation distribution (Anderson, 1982). Unlike other methods of estimating home range, neither of these makes any *a priori* assumption as to shape of home range. The utilisation distribution is also relatively independent of the number of locations.

Results

In his original definition of home range, Burt (1943) recognised that size and location of the home range of an individual need not be invariable throughout the life of that individual. The home range of a migratory animal in summer is quite distinct from that in winter. Furthermore, no fixed time span is incorporated into the definition of home range. An individual's home range can be estimated for any specified duration.

In the study carried out on Kinchega and Tandou home ranges were calculated for three discrete durations (Table 7.1). These were: (i) a 'night range' calculated from all locations of an individual recorded during a single night; (ii) a 'seasonal range' calculated from all locations of an individual recorded during a three-week period each season; and (iii), an 'aggregate range' calculated from all locations of an individual recorded during the study.

Between December 1980 and February 1982 red and western grey kangaroos were not wide-ranging. The mean aggregate range was less than 8.0 km² for both species (Table 7.1). Many home ranges overlapped.

The distance between the arithmetic centres or 'the centres of activity' (Hayne, 1949) of two home ranges of the same individual recorded at different times was defined as the shift in location of the home range. The average shift between consecutive night ranges was 1,060 m for red kangaroos and 900 m for western grey kangaroos. If kangaroos continually shifted their home range at this rate they would be extensively nomadic, moving hundreds of kilometres each year. Yet a mean shift in the location of the seasonal range of less than 1,400 m for reds and 900 m for western greys over a three-month period (Table 7.1) indicated conclusively that both species were sedentary rather than nomadic.

Telemetered kangaroos were also radiotracked by aircraft during the day. Unlike the night locations which included both feeding and resting sites the daytime locations were resting sites only. Each kangaroo was located once on each of three consecutive days each season. Home ranges were also calculated from these data (Table 7.2). For red kangaroos the area of the aggregrate range was similar to that obtained at night by radiotracking from the fixed tracking stations (Table 7.1), but for western grey kangaroos it was considerably less. This indicates that during the day red kangaroos used resting sites that were as widely dispersed as the areas in which they fed at night; possibly they rested where they fed. On the other hand western grey kangaroos ranged over a similar sized area as red kangaroos to feed (7-8

Table 7.1. *Estimates of the mobility of kangaroos derived from radiotracking from fixed stations. Values are means* ± *SE. CP: estimate based on the minimum convex polygon (Southwood, 1966). MAP: estimate based on the utilisation distribution (Anderson, 1982).*

	Red kangaroos	Western grey kangaroos
Area of home range (km²)	n	n
night range (CP)	1.58 ± 0.28 (61)	1.11 ± 0.17 (75)
seasonal range (CP)	2.46 ± 0.21 (63)	2.37 ± 0.20 (75)
seasonal range (MAP)	2.15 ± 0.21 (64)	2.13 ± 0.21 (75)
aggregate range (CP)	7.74 ± 0.90 (21)	6.92 ± 0.77 (23)
aggregate range (MAP)	4.63 ± 0.49 (21)	3.94 ± 0.54 (23)
Shift in home range (km)		
night range	1.06 ± 0.11 (47)	0.90 ± 0.08 (62)
seasonal range	1.32 ± 0.16 (44)	0.83 ± 0.09 (53)

km²), but by mid-morning they had returned to resting sites that were not so widely dispersed.

The centre of each individual's three daytime resting locations was calculated for each season. The distance between these central locations did not increase as a function of time (Table 7.3). That is, resting sites separated in time by 12, 9 or 6 months were located no farther apart than those separated by only 3 months. Clearly, individuals of both species were not nomadic.

Three-factor analysis of variance of data obtained by radiotracking kangaroos from fixed tracking stations showed that: (1) on average, red kangaroos had larger home ranges than western grey kangaroos (7.7 and 6.9 km²

Table 7.2. *Estimates of the mobility of kangaroos derived from radiotracking from fixed-wing aircraft. Values are means ± SE. CP: estimate based on the minimum convex polygon (Southwood, 1966). MAP: estimate based on the utilisation distribution (Anderson, 1982).*

	Red kangaroos		Western grey kangaroos	
Area of home range (km²)		*n*		*n*
seasonal range (CP)	0.32 ± 0.06	(98)	0.16 ± 0.03	(94)
aggregate range (CP)	7.28 ± 1.57	(23)	3.51 ± 0.78	(22)
aggregate range (MAP)	5.99 ± 0.84	(23)	2.69 ± 0.59	(22)
Shift in home range (km)				
seasonal range	1.79 ± 0.20	(85)	1.05 ± 0.12	(81)

Table 7.3. *The mean shift in location of the home range (km).*

From	To			
	April 1981	June 1981	Sept. 1981	Dec. 1981
Red Kangaroos				
Dec. 1980	2.098	2.002	1.280	0.896
April 1981		2.138	2.536	1.908
June 1981			1.379	1.739
Sept. 1981				1.540
Western Grey Kangaroos				
Dec. 1980	0.849	1.041	0.784	0.724
April 1981		0.916	0.796	0.659
June 1981			0.896	0.744
Sept. 1981				0.479

respectively); (ii) males had larger home ranges than females (8.1 and 6.5 km² respectively); and (iii) kangaroos in Tandou had home ranges that were marginally larger than those of kangaroos in Kinchega (7.7 and 7.1 km² respectively). These relationships held for both methods used to calculate home range. An identical analysis of home range calculated from the daytime resting sites located by aerial radiotracking gave similar results.

The difference in the home range of red and western grey kangaroos may reflect differences in their diet. Red kangaroos are more selective towards green forage (Fig. 9.1) and consequently may need to range further afield to find it. Differences between males and females in the size of their home range may be related to differences in their body size. Alternatively, the larger home range of males may be the result of them wandering further than females when seeking mates. The differences in size of the home range between kangaroos on Kinchega and those on Tandou probably reflect differences in vegetation each side of the boundary fence. These differences may not always

Table 7.4. *Correlation coefficients between size of home range and weather and pasture variables.*

Variable	Red kangaroo	Western grey kangaroo
Minimum temperature		
for preceding month	0.96**	−0.27
for preceding 3 months	0.85	0.34
Maximum temperature		
for preceding month	0.82	−0.54
for preceding 3 months	0.95*	0.09
Solar radiation		
for preceding month	0.83	−0.69
for preceding 3 months	0.99**	−0.59
Evaporation		
for preceding month	0.94*	−0.47
for preceding 3 months	0.99**	−0.08
Rainfall		
for preceding month	−0.60	0.68
for preceding 3 months	−0.86	0.14
Pasture biomass	0.54	−0.86
Pasture growth for preceding 3 months	−0.77	−0.13

** Significant at $P = 0.05$
* Significant at $P = 0.10$

be apparent as, although the pasture biomass in Kinchega was on average 47% greater than that in Tandou during the period of radiotracking, the long term averages are similar (Chapter 4).

Changes in size of home range over time were examined by simple linear correlations with weather and pasture growth and biomass (Table 7.4). Although the data used in these computations were limited, and many of the correlations were not significant, the results yield some useful insights. They suggest that the determinants of home range differ between red and western grey kangaroos.

For red kangaroos, area of home range was positively correlated with temperature, solar radiation and evaporation, and negatively correlated with rainfall and the growth of pasture. In essence, home range size increased with increasing aridity. Since the diet of red kangaroos consists predominately of succulent herbs and grasses (Chippendale, 1962) it is easy to envisage how hot dry weather and a decrease in pasture growth could compel them to expand their home range in search of remnants of green forage. A corollary of the relationship between home range and aridity is that in the arid areas of Australia home ranges of red kangaroos should be larger than those of kangaroos adjacent to the Darling River floodplain. This has yet to be confirmed.

For western grey kangaroos, the most highly correlated variable was pasture biomass ($r = -0.86$). Western grey kangaroos did not increase their mobility in relation to aridity or a decrease in pasture growth. This differs from red kangaroos and indicates that western grey kangaroos are not as dependent on fresh green herbage as red kangaroos. That inference is strengthened by a comparision between species in fibre content of their food (Fig. 9.1), the western grey kangaroo being more efficient at digesting fibre (Prince, 1976).

7.4   Use of habitats by red and western grey kangaroos

Work carried out on Kinchega and Tandou aimed first to compare habitat utilisation between red and western grey kangaroos; and second, to determine whether there were any changes in habitat utilisation over time. Habitats were classified according to landform (floodplains or sandplains) and vegetation growth form (woodland, shrubland or open plains). The alluvial soils of the floodplains are very different to the aeolian soils of the sandplains (Chapter 2) and pastures on each soil type respond differently (in terms of germination and growth) to rainfall and weather (Chapters 3 & 4). Dispersion of kangaroos was examined in relation to the distribution of each landform and vegetation type.

Methods

Between March 1980 and December 1981 the habitats occupied by each species of kangaroo were recorded in Kinchega National Park and Tandou by observation from an aircraft during the quarterly aerial surveys described in Chapter 8. Two replicate surveys on consecutive days were flown each quarter. Kangaroos within the prescribed transect were counted and assigned to a particular habitat according to the soil and vegetation type they occupied. The total number of each species seen in each habitat was tallied. The different habitats were easily discernible from the air, and generally the boundaries between them were discrete.

Results

The dispersion of kangaroos changed substantially; usually kangaroos were widely dispersed, but when pastures were scarce they aggregated on areas of remnant pastures in depressions and along the edges of receding lakes and creeks. Data from each of the eight quarterly aerial surveys were analysed to determine whether western grey kangaroos utilised the various habitats differently from red kangaroos.

Sandplains were used by a greater proportion of the red kangaroo population than of the western grey population. Correspondingly, floodplains were used relatively more by western greys than by reds. Red kangaroos used open plains to a greater extent than did western grey kangaroos. Correspondingly, areas of woodland were used relatively more by western greys than red kangaroos.

Both red and western grey kangaroos showed significant changes over time in the use of the two landforms, indicating there had been substantial movements from one to the other. There were no significant changes in use of the various vegetation types by either species. At no time during the study was there any collective movement between woodland, shrubland, or open plains. Therefore the movements between landforms were independent of the vegetation types. Changes in the use of the two landforms were not seasonal but paralleled changes in overall pasture biomass (Fig. 7.7). When the environment became dry and pasture growth declined, kangaroos moved onto the floodplains. It is unlikely that they moved onto these low-lying areas to drink, as surface water was not present in many areas. More likely, kangaroos were attracted there to graze the last remnants of palatable pasture. There was no difference in the pasture biomass on each landform (Chapter 4), but the different moisture-retaining properties of the contrasting soils of each landform affected the survival of pastures differently (Chapter 3). Although not verified empirically, it appeared that pastures on the alluvial soils of the floodplains stayed palatable longer than those on the aeolian soils of the sandplains, especially on areas adjacent to creeks, billabongs and depressions.

Although floodplains appeared to be refuges for kangaroos when conditions were dry, pastures on the heavy-textured soils did not regenerate as rapidly or as profusely following light rain as those on the lighter-textured soils (Chapter 4). Consequently, kangaroos moved onto the sandplains after rain to take advantage of the rapid growth of pastures on these lighter-textured soils. Radiotracking (Priddel, unpubl. data) showed that red kangaroos moved to the sandplains immediately following rain, whereas western greys moved two to ten weeks later.

Not all kangaroos took part in these movements. To account for all observed changes in the dispersion of kangaroos, only 40% of the population need have

Fig. 7.7. Relative use of the two landforms — floodplains and sandplains — by kangaroos on Kinchega. Preference for sandplains calculated as the ratio of the number of kangaroos counted on sandplains relative to the number counted on floodplains. Points are means of two replicate surveys. Estimates of pasture biomass are taken from Chapter 4. Pasture biomass was not measured prior to August 1980.

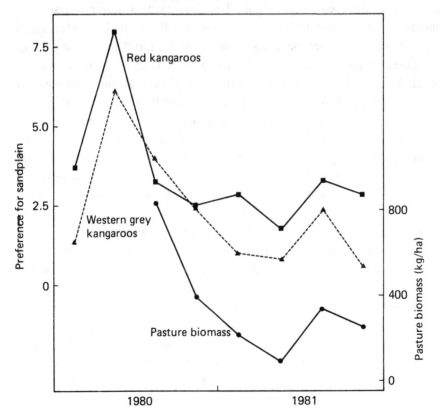

moved. Johnson & Bayliss (1981) identified a differential pattern of movement of red kangaroos on Kinchega relative to age, sex, and reproductive condition. They found large males and lactating females concentrated on the most favourable pasture, whereas sub-adults, small males and non-lactating females were more thinly dispersed elsewhere. Movements of kangaroos between landforms in response to changes in pasture productivity have been reported also by Low *et al.* (1981). They reported (p. 32) that the floodplains and gilgai plains, comprising self-mulching grey clays, received their greatest use during drought. Correspondingly, areas of mulga shrubland, comprising red-earth soils, received their greatest use when pastures were abundant following rain.

7.5   Discussion

Extrapolation of information gathered at Kinchega to other areas of Australia must take into account the abundance of surface water, the interspersion of different landforms with contrasting soil types, and barriers to movement at Kinchega.

A permanent supply of surface water is typical of most of the Australian rangelands since European settlement. The few natural water-courses are supplemented by numerous bores and earth dams. Without this water much of the rangelands could not be stocked with sheep. Land more than 5 km from permanent water is unsuitable for them (Soil Conservation Service of New South Wales). Only in remote arid areas of central and mid-western Australia are red kangaroos restricted by availability of permanent water. Water is rarely in short supply for the majority of red and western grey kangaroos and the major determinant of their dispersion or movements is the quantity and quality of pastures (Newsome, 1965b). Many bores, dams, and some natural water courses retain water during drought but most pastures wither and die, imposing a shortage of food on kangaroos and domestic stock alike.

Both red and western grey kangaroos moved between sandplains and floodplains in response to pasture growth. The most beneficial home range would therefore be one that contained interspersed areas of contrasting soil types. Such a home range would enable kangaroos to graze the residual palatable pasture from heavy-textured soils as the environment became drier, and yet enable them to take full advantage of any emergent growth of pastures on light-textured soils immediately following drought-breaking rains. Kangaroos inhabiting extensive tracts of a single landform may behave quite differently. Without access to a mosaic of land systems locally it is likely that they would range over much larger distances. This has yet to be verified.

By surveying the distribution of eastern and western grey kangaroos in New South Wales, Scott-Kemmis (1979) similarly concluded that the heterogeneity of land systems was important in determining the distribution of these animals. He found the interspersion of open plains of heavier-textured soils with undulating areas of lighter-textured soils provided favourable habitat conditions. In contrast, extensive tracts of a single vegetation, soil or landform type did not support high densities. Kirkpatrick (1967), Bell (1973), and Grant (1973) highlighted the eastern grey kangaroo's preference for areas containing heterogeneous habitats, particularly the transitional areas between woodland and open plains. Similarly, Frith (1964) suggested that red kangaroos in inland New South Wales favoured grasslands close to areas of woodland. Hence, if areas are to be set aside for the protection and conservation of red or grey kangaroos, then they should contain a diversity of habitats and soil types. In this respect the selection of Kinchega as an arid-zone National Park was an appropriate choice.

Kinchega National Park is enclosed by a 'macropod-deterrent' fence to the south and west, by Menindee Lake to the north, and by the Darling River to the east. The fence was erected to prevent the dispersal of kangaroos from the Park. The fence is not an effective barrier. Several tagged kangaroos had home ranges which included areas on both sides of the boundary fence. These individuals regularly moved back and forth across the fence. Most kangaroos in the Park remained within home ranges that were small (< 8 km^2) in relation to the size of the Park (440 km^2). Many of these home ranges did not extend to the boundary fence and presumably many kangaroos have never encountered the fence. Although the density of kangaroos inside the Park is higher than that in nearby areas outside (Chapter 8) the difference is not due to the fence restricting the kangaroo's ability to disperse. Rather, it is due to the absence of sheep grazing within the Park (Chapter 8). The fence is only important in that it prevents sheep from entering.

Most red and western grey kangaroos on Kinchega and Tandou were sedentary. The advantage of maintaining a relatively sedentary existence is that the location of shelter and erratic supplies of food and water are all known. Kangaroos are then readily able to utilise these resources without necessarily spending time to search for them. Red kangaroos are known to make regular excursions between their feeding, resting and watering sites (Bailey, 1967; Priddel, unpubl. data) and in Kinchega, both species were observed to journey several kilometres along distinct, well worn routes to drink from the Darling River.

Collars retrieved from dead animals on both Kinchega and Tandou during the drought of 1982-3 provided stark evidence that many kangaroos did not move off in search of green pasture, as popularly believed, but instead remained and died within their home range. In time of drought kangaroos have depleted fat reserves and poor body condition (Chapter 9). Moving hundreds of kilometres in search of green pastures would place additional demands on their reserves. If they were unsuccessful in finding food and water they would die sooner than if they had remained within their home range.

High mobility of kangaroos is not necessary to account for the changes observed in their dispersion. Such changes can be attributed to local aggregation, recruitment and high mortality. As extensive areas become denuded of pasture during drought, kangaroos move within their home range to aggregate on areas of remnant pasture. The dramatic increase in kangaroo density in these areas creates a false impression of high population levels. As pastures wither, most kangaroos remain and many subsequently die. This high mortality causes a sudden decrease in kangaroo density which is often interpreted incorrectly as a mass exodus of animals from the area rather than a high rate of natural mortality.

7.6   Conclusions

Most red and western grey kangaroos on both Kinchega and Tandou were sedentary, being confined to home ranges of less than 8 km². For reasons unknown, a few individuals of both species occasionally ranged widely. Red kangaroos increased their home range when pasture growth declined during times of increasing aridity. Western grey kangaroos increased their home range only when total pasture biomass diminished. Both species increased their use of floodplains as conditions became drier, but returned to the sandplains after rain. Red kangaroos moved first, western greys followed later. The fence surrounding Kinchega was not an effective barrier against kangaroo dispersal.

8

Kangaroo dynamics

PETER BAYLISS

8.1   Introduction

This chapter examines the relationship between the population dynamics of kangaroos and their food supply in an arid rangelands environment. The relationship is called the numerical response (Solomon, 1949) and was introduced to quantitative ecology by Holling (1959, 1961) to model one component of a predator-prey system. Trends in numbers of red and western grey kangaroos on Kinchega National Park and on surrounding properties were monitored every three months between 1973 and early 1984 by standardised aerial survey. Rates of increase of kangaroos were plotted as a function of rainfall and pasture biomass. Rainfall is the dominant influence on an arid zone grazing system (Noy-Meir, 1973) affecting the rate of increase of both the plants and the animals that eat them. However while rainfall is probably the ultimate factor determining rate of increase of kangaroos, the biomass of pasture available to the kangaroos as food is the proximate factor. Pasture biomass was measured on Kinchega and on an adjoining sheep station, Tandou, at three-monthly intervals coinciding with the aerial surveys between 1980 and early 1984 (Chapter 4).

There are few studies of the population dynamics of large mammalian herbivores, probably because most grazing systems are stable and so exhibit little dynamic behaviour. There are virtually no grazing studies that integrate the dynamic interaction of plants and animals, the exception being Sinclair's (1977) study on the African buffalo. The quantitative excursion into the population dynamics of kangaroos in this chapter is enhanced rather than hindered by the vagaries of rainfall in the arid zone. The dynamics of a system can be studied only when the system is displaced from its hypothetical equilibrium or static state: to study change we need change. In the arid zone, the rapid and marked changes in plant and animal biomass allow the population dynamics of large mammalian herbivores and the plants they eat to be elucidated in a very short period of time. The years 1973 to 1976 received exceptionally high rainfalls providing kangaroos with a superabundance of food. Annual rainfalls between 1977 and 1981 were close to average, but in 1982 rainfall was only 50% of the long-term average. Hence kangaroos during this study experienced times of plenty, average conditions, and times of drought.

8.2   Trends in numbers

8.21   *Methods*

Kangaroos were monitored on Kinchega and adjoining sheep stations by standardised aerial surveys at three-monthly intervals for ten years. The survey design is discussed in detail by Bayliss (1980, 1985a, b). Both species were counted on systematic sample transects by two observers, one on either side of a Cessna 206. The boundaries of the transects were demarcated by streamers attached to each wing strut. Each survey sampled 39.9 km² of

Fig. 8.1. Observing kangaroos from the air is relatively easy in open country. These are red kangaroos.

country (9% of the Park), beginning soon after first light when kangaroos were most active and visible. Two survey replicates were flown on the Park on consecutive days. On a third day the sheep properties were surveyed once. Aerial surveys of Tandou sheep station (south of the Park) were conducted from 1978 in conjunction with surveys of the Park and surrounding sheep properties. Two survey replicates were flown in the early morning on consecutive days sampling 38.5 km^2 of country, a sampling intensity of 9%.

Estimates of density

Two different estimates of density are used in subsequent analysis and discussion. The first, a yearly average density estimate, is used in the comparison of changes in rate of increase with rainfall over ten years. The second, a seasonal estimate of density, is used in the comparison of rate of increase with pasture biomass for the four years that such data are available. Their derivation is discussed below.

Yearly estimates of density

Kangaroos counted by both observers on all transects of a survey were added for each species in each location (Kinchega, Tandou, other properties). Indices of density derived by aerial survey are negatively biased in that they underestimate the true population (Caughley, Sinclair & Scott-Kemmis, 1976). Here these counts were corrected for visibility bias due to ambient temperature (Bayliss & Giles, 1985), cloud cover (Short & Bayliss, 1985) and vegetation cover (see Appendix 8.2). Yearly averages were calculated as the mean estimate of all surveys flown during the year, divided by the constant area sampled at each survey when an estimate of absolute density was required.

Seasonal estimates of density

The three-monthly estimates of density for both species, corrected for visibility biases, were examined in detail on Kinchega and Tandou between 1980 and early 1984. During this period concurrent estimates of the standing biomass of pasture were available for both Kinchega and Tandou.

Exponential rates of increase (r)

The estimates of density were transformed to natural logarithms to calculate exponential rates of increase. The exponential rates of increase between consecutive years (r) was calculated as the logged density estimate in one year minus the logged density estimate of the previous year. The seasonal exponential rates of increase (r) for the period 1980 to 1984 were calculated by taking the log of the corrected mean density estimate in one season and subtracting the equivalent estimate from the same season of the previous year. This gives a yearly instantaneous rate centred between the two estimates of density. Exponential rates of increase for each three-month period are not fully independent because they are calculated over periods that overlap. They correspond to estimates of pasture biomass for the same period (Chapter 4).

8.22   *Results and discussion*

Densities of kangaroos between 1973 to 1984

Figure 8.2 plots the yearly trends in red and western grey kangaroo densities

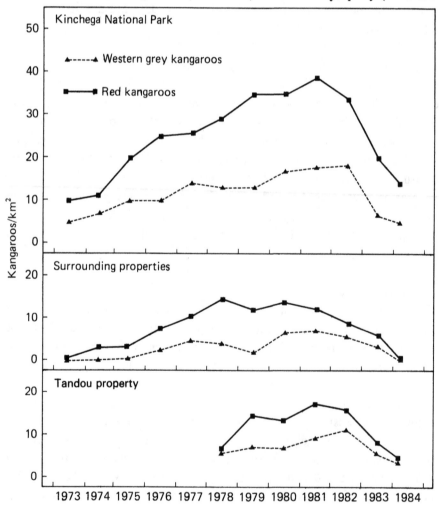

Fig. 8.2. The yearly trend in red and western grey kangaroo densities on: Kinchega National Park (1973-1984), sheep properties abutting the Park west of the Darling River (1973-1984) and Tandou property (1978-1984).

on Kinchega and on sheep properties abutting the Park between 1973 and 1984, and on Tandou between 1978 and 1984. Changes in kangaroo numbers reflect years of above average rainfall in the early to mid 1970s and drought in 1982-3.

In 1973 there were approximately 10 reds and 5 western greys per square kilometre inside the Park, and 2 reds and 1 western grey per square kilometre outside. These low densities presumably reflect the severe drought in 1965-7. Between 1973 and 1977 all populations averaged high annual rates of increase and doubled in density twice during that period (Fig. 8.2). Between 1977 and 1982 average rate of increase was close to zero for all populations (Bayliss, 1985a). The shift from low to high densities, and the persistence of these high densities for five years, reflected the atypically high rainfalls between 1973 and 1976 followed by adequate rainfalls thereafter. In 1974 the study area received three times its annual average rainfall (739 mm as against 236 mm). The collapse of these high densities occurred between late 1982 and early 1983 during severe drought. In 1982 annual rainfall was half the annual average (126 mm). All populations were approximately halved by the drought and apparently continued to decline into early 1984 although drought-breaking rains occurred in March 1983. The population trends since 1973 for kangaroos on properties abutting the park mirror those for kangaroos inside the park. There is clearly a strong general relationship between annual rainfall and rate of increase of kangaroos inside and outside the Park.

8.3   Comparison of national park with sheep stations

The ratio of reds to western greys inside the Park was 3.2 to 1, and that for Tandou sheep station 2.3 to 1, perhaps reflecting more woodland habitat for western greys on Tandou. Kangaroo densities inside the Park were on average twice that of Tandou. Despite these differences kangaroo numbers inside and outside the Park appeared to move in concert. Numbers on Kinchega and Tandou built gradually to a peak in 1980 and 1981. Kangaroos on the other sheep stations appeared to plateau between 1978 and 1980. On all areas studied kangaroo numbers declined markedly in 1982 and 1983 as a consequence of drought (Fig. 8.2). Bayliss (1985a) compared average rates of increase of red and western grey kangaroos on Kinchega and some adjoining sheep stations between 1973 and 1981 and found no statistically significant difference, although the observed rates for red kangaroos were generally a little higher than those of western greys in both areas. Thus we have an apparent paradox in which the density of kangaroos (both species combined) on Kinchega National Park is about double that on the adjoining sheep station whereas the dynamics per head are much the same in the two areas. We seek a mechanism or agent that acts to produce a sharp contrast in long-term mean density while having little or no effect on short-term rates of increase. Predation, parasites and diseases (Arundel, Beveridge & Presidente, 1979) were insignificant in the study area and hence are unlikely to provide a contrast between the two areas. Availability of cover and water is more similar than different between the two areas. The anomaly is more likely to reflect the influence of harvesting of kangaroos (absent on Kinchega but not on Tandou), or of dispersal (constrained on Kinchega but not on Tandou), or of an interaction with sheep (absent on Kinchega but present on Tandou). These influences are examined below.

8.31   *Harvesting*

Annual harvests (numbers/km^2) of red and western grey kangaroos on Tandou property are compared with their average yearly density (corrected for temperature and cloud cover effects) in Fig. 8.3 between 1978 and 1983. Illegal shooting cannot be assessed but was assumed to be insignificant. Commercial harvests in years previous to 1978 were less than 1% of the total kangaroo population (Bayliss, 1980). The offtake of red kangaroos was less than 8% of the population between 1978 and 1981; however this increased substantially to 33% in 1982 and 17% in 1983. Western greys showed a similar trend, their offtake being less than 7% between 1978 and 1981 but increasing substantially to 16% in 1982 and 27% in 1983. These increases in offtake reflect both a high demand for kangaroo skins and the increased need to reduce competition with sheep for food during drought. Harvesting appeared to have an insignificant effect in causing differential densities between locations for the period 1973 to 1984. Although commercial harvesting was substantial during the drought of 1982, the average population declines during this drought for kangaroos on Kinchega and Tandou were similar. Perhaps some animals that would otherwise have died in the drought were harvested instead, thus increasing the chances of survival (more scarce food per head) of the remainder. Whether kangaroo hunting reduces non-harvest mortality remains to be tested.

8.32   *Dispersal and social behaviour*

Numerical responses acting through fecundity and survival rates may be standard for many species but dispersal may be more important for others (Crawley, 1975). Herbivores may respond quickly by emigration to reduced food supply and the disadvantages of overcrowding. Conversely, a numerical response through immigration can increase herbivore numbers in areas of high food supply more rapidly than would be possible by reproduction.

Some forms of territoriality and social behaviour may reduce a population's rate of increase to avoid the disadvantages of overcrowding (Birch, 1960). Hence the effects of herbivore density on rate of increase may be two-fold: one intrinsic to the population (self-regulation) involving social behaviour or dispersal, the other extrinsic to the population where the amount of available food is determined by the density of the herbivores eating it. Krebs (1971) described a 'fence-effect' for a population of voles with restricted dispersal, whereby increased aggressive behaviour increased mortality and decreased natality. The population of kangaroos at Kinchega was surrounded by a fence but it was not an effective barrier against their movement. Some kangaroos moved freely across the southern boundary (Chapter 7), both jumping the fence and utilising holes pushed through it. However, these movements were not regarded as large-scale dispersal but as movements within a home range that overlapped a leaky part of the fence. Caughley & Krebs (1983) suggest that for large mammalian herbivores self-regulation is unlikely to evolve unless the maximum rate of increase of a population with a stable age distribution exceeds $r = 0.45$, this value corresponding to an average population body weight of about 30 kg. Since kangaroos have a body weight and intrinsic rate of increase near those values their dynamics may lie on a cusp.

Fig. 8.3. A comparison of the density of red and western grey kangaroos and their harvest rates on Tandou property between 1978 and 1983.

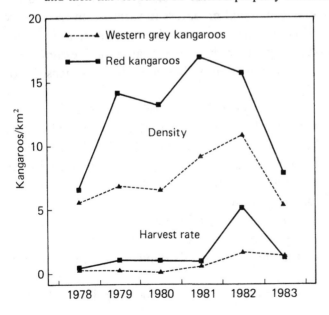

8.33   *Sheep*

The data suggest that kangaroo harvesting and large-scale dispersal cannot account for the large difference in density between Kinchega and Tandou. Bayliss (1985a) suggested that large-mammal grazing pressure was similar in the two study areas. A reasonable hypothesis is that kangaroos and sheep shared the same food supply, and that over time this suppressed kangaroo densities on Tandou. In favour of such a hypothesis are the many dietary studies of kangaroos and sheep which show more similarity in the plants they eat than differences, especially after rain (Chapter 5; Griffiths & Barker, 1966; Storr, 1968; Ellis, Russell, Dawson & Harrop, 1977). This hypothesis is examined in detail in Chapter 10.

8.4   Mortality of kangaroos during drought

The reduction in the density of kangaroos on Kinchega over the summer of 1982-3 was estimated from the mean drop in seasonal indices for the three 12-month periods commencing March, June and September 1982. Red kangaroos declined by 41% and western grey kangaroos by 62%, an average drop in numbers of 52%. Similarly, reds and western grey kangaroos declined on Tandou by 52% and 54% respectively, representing an average drop in numbers of 53%.

8.5   Numerical response

Holling (1959) examined the numerical response of small mammals preying upon European pine sawfly, and Morris *et al.* (1958) examined the numerical response of bird and mammal predators to variation in the density of the spruce budworm. Although these two decades-old studies are classics for quantitative population ecology, there has been no attempt to examine the numerical response of a large mammalian herbivore to different densities of food supply.

This section examines the components of the relationship between the dynamics of each kangaroo population (measured by *r*) and their food supply, as indexed first by rainfall and then by pasture biomass (kg/ha of dry weight). The relationship between rainfall and rate of increase is examined for its potential predictive power because rainfall data are readily available but measurements of pasture biomass are not.

The approach used here is one of consonant modelling, whereby the model constants correspond to biological processes. Consonant modelling has its best examples in the field of invertebrate population dynamics (e.g. Crawley, 1975; Gilbert *et al.*, 1976).

Kangaroo populations may respond to changes in food supply by adjusting their rates of reproduction, survival, or both: the rate of increase summarises the response. A numerical response model therefore describes the essential components of a population's dynamics in relation to food supply (Gause, 1934; Ivlev, 1961; Leslie, 1966; Tanner, 1975). Many of the models used to describe the functional response (see Chapter 6) by Rosenweig (1971), Holling (1959, 1961), and Noy-Meir (1978) may also apply to the numerical response (May, 1973). Ivlev's (1961) inverted exponential function was used because it was found to be least sensitive to outlying data points.

The numerical response function in Ivlev form is:

$$r = -a + c\,(1 - e^{-dV})$$

where r is the yearly exponential rate of increase of kangaroos; a is the maximum rate of decrease in the absence of food; c is the rate at which a is progressively ameliorated by increasing the amount of food until the herbivore is satiated; d is a measure of the demographic efficiency of the kangaroo population, or its ability to multiply when V is sparse (the lower the numerical value of d, the lower is demographic efficiency); and V is plant biomass. The asymptotic or maximum rate of increase (r_m) (Birch, 1948; Caughley & Birch, 1971; Bayliss, 1985b) equals $c - a$.

The approach to fitting Ivlev's model to the rate of increase, rainfall and plant biomass data for all populations is of necessity *ad hoc*. There were three steps. First, the apparent asymptote for each population was fixed at the highest rate of increase measured. These r_m values (Table 8.1) fall within the range of those estimated by J. Caughley, Bayliss & Giles (1984) and Bayliss (1985a, b). Second, a linear regression was fitted to the remaining points. The regression intercept was used to estimate a, and c was calculated as ($r_m - a$). All linear regressions used to calculate a for each kangaroo population were significant, explaining approximately 40% of the variability of the data. The maximum rates of decrease are greater than those estimated by the above authors. Earlier data lacked a major decline in kangaroo numbers during drought. The demographic efficiency (d) of all populations was estimated as the slope of the regression through the origin of $-\ln\,[1 - (r + a)/c]$ on V, where V can be either the standing pasture biomass or the index to that provided by annual rainfall lagged by six months.

8.51   *Rainfall and rate of increase*

Figure 8.4 shows the relationship between lagged annual rainfall and the rate of increase of red and western grey kangaroos on Kinchega National Park. The data span ten years (1973 to 1983) and includes years of exceptionally high rainfall and drought years. Only the numerical response curve for kangaroos on Kinchega National Park is shown because Bayliss

(1985a, b) showed that the reactions of all populations to rainfall between 1973 and 1981 were similar. The replicate three-monthly surveys on the Park provided a more stable index of abundance and hence rate of increase than the single-survey estimate of the surrounding properties.

The average time lag between rainfall and the detected response by aerial survey is six months (Bayliss, 1985a; Caughley, Bayliss & Giles, 1985). Annual rainfall and yearly density estimates centre on the 1st of July, and the annual instantaneous rate of increase is centred on the 1st of January. The shortness of the time lag implies a survival response and not a breeding response. The greatest mortality of kangaroos in times of food shortages is in pouch young during advanced lactation (Newsome, 1965c, 1966; Kirkpatrick & McEvoy, 1966; Poole, 1973; Russell, 1974). If rainfall is high and food abundant, juvenile survival is high and is detected as a positive change in rate of increase six months later.

Covariance analysis of the linear regressions of rate of increase on rainfall (up to the asymptotic level) between species showed that the demographic efficiency and maximum rate of decrease of each were not significantly different, reinforcing the conclusion that there is no important difference in the population dynamics of the two species. Both species probably share the same food supply (Chapter 5) and differences in survival and fecundity between the species are likely to be trivial overall (Bayliss, 1985a, b). Although

Table 8.1. *The maximum rate of increase ($r_m = c - a$), maximum rate of decrease (a), demographic efficiency (d) and c values estimated for Ivlev's (1961) numerical response model using rainfall and pasture biomass as indices of kangaroo food supply.*

Index of food supply	Location	Species	r_m	a	c	d	$P*$
Rainfall	Kinchega						
		red	0.57	0.57	1.14	−0.004	0.05
		w. grey	0.35	0.69	1.04	−0.005	0.05
Pasture biomass	Kinchega						
		red	0.34	0.80	1.14	−0.007	0.01
		w. grey	0.42	1.01	1.43	−0.007	0.01
	Tandou						
		red	0.44	0.73	1.17	−0.007	0.05
		w. grey	0.66	0.64	1.30	−0.007	0.05

* significance of the linear regression used to calculate maximum rate of decrease.

red kangaroos should have a higher demographic efficiency than western grey kangaroos in variable arid and semi-arid environments (Chapter 9), the results from Kinchega and Tandou suggest that the continuous birth rate, higher turnover of pouch young, and marginally more rapid breeding response to rainfall of red kangaroos did not advantage them greatly relative to western grey kangaroos during the ten years of this study. The advantages of continuous (reds) and seasonal (western greys) breeding strategies were probably equally weighted in this environment where similar amounts of rainfall occur in all seasons.

The annual rainfall at which rate of increase is zero is similar for both species (red kangaroos: 198 mm; western grey kangaroos: 217 mm) and to the long term annual average (236 mm). J. Caughley *et al.* (1984) found rate of increase to be zero at similar rainfall for both species of kangaroos in western New South Wales (200 mm and 215 mm respectively). These estimates are approximatley 40% more for each species than those obtained by Bayliss (1985b) using a Michaelis-Menten model (Noy-Meir, 1978). More than likely the difference is explained by poor extrapolation of earlier data (analysis was for the period 1973-81) which lacked rate of decrease estimates from a severe and prolonged drought.

The simplicity of the relationship between rainfall and rate of increase, and the ready availability of rainfall figures, provides a convenient predictive tool for kangaroo management. If trends in kangaroo numbers can be predicted six months ahead from rainfall data then culling quotas may be set with greater assurance.

Fig. 8.4. The numerical response function of red and western grey kangaroos on Kinchega National Park in relation to rainfall with a six-month lag (red kangaroos, $r = -0.57 + 1.14(1 - e^{-0.004R})$; western grey kangaroos, $r = -0.69 + 1.04(1 - e^{-0.005R})$. Open symbols represent data not used in calculating the functions.

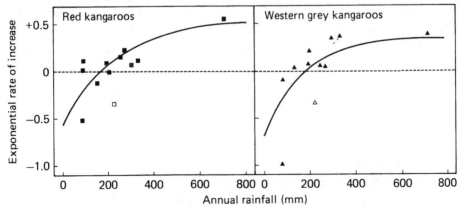

8.52   *Food supply and rate of increase*

The exponential rates of increase (r on a yearly basis) of kangaroos on Kinchega and Tandou for each three-monthly interval between 1980 and 1983 were plotted against pasture biomass. Rates of increase were related best to pasture biomass estimates (kg/ha dry weight) with no time lag. Figure 8.5 plots the numerical response functions in Ivlev (1961) form for all populations. Three of the four data sets were significantly curvilinear when tested by polynomial regression (Zar, 1974) confirming the *a priori* assumption that rate of increase asymptotes at high pasture biomasses.

The 'equilibrium' vegetation levels where rate of increase is zero (Caughley, 1976a) varied between 95 and 125 kg/ha for western greys and 140-180 kg/ha for red kangaroos. Pasture biomass between 1980 and 1984 was below these levels for 29% and 41% of the time for each species respectively (Chapter 4). These levels are not the equilibrium point as conventionally interpreted from these models (see Chapter 10). Kangaroos are not 'stable' at these 'equilibrium' levels because of large yearly variations in rainfall (and hence plant biomass) in arid zones. The equilibrium level is a function not only of extrinsic density effects (via negative feedback through the functional and numerical responses) but also of weather and other grazing components in the system unaccounted for (e.g. invertebrates). The grazing system is essentially stochastic.

Covariance analysis of the linear regressions of rate of increase on plant biomass (up to the level of satiation) shows that the demographic efficiency and maximum rate of decrease of all populations were not significantly different. A similar result was obtained in the previous section when rate of increase was regressed against rainfall.

A critical assumption of the numerical response model is that kangaroo density has no instantaneous effect on their rate of increase, although it may be affected by previous densities reducing food supply. Bayliss (1985a) attempted to separate the independent effects of rate of increase on rainfall and kangaroo density but the analysis failed because of a coincidental trend between the two. Similarly, the independent effects of plant biomass and kangaroo density could not be ascertained in this study. However, a strong relationship between rate of increase and plant biomass obtained here argues against the importance of intrinsic density effects.

8.53   *Parameters of the model*

Maximum rate of decrease

The numerical response function using rainfall as an index of food supply produced maximum rates of decrease less than those derived by using pasture biomass (Table 8.1). However, the maximum rate of decrease calculated from yearly indices would be less than that calculated from the quarterly indices which reflect extreme and real fluctuations in numbers within a year: these fluctuations are smoothed out when averaged over one year.

There are two anomalous negative rates of increase at high levels of food in all numerical responses plotted as a function of pasture biomass, and these were not used in any of the analyses (Fig. 8.5). Kangaroos appeared to decline despite high levels of plant biomass resulting from drought-breaking rains in March 1983.

Kangaroos on Kinchega and Tandou registered an average rate of decline of -0.40 despite pastures of 800 kg/ha. Similar trends were obtained for red

Fig. 8.5. The numerical response function of red and western grey kangaroos on both Kinchega National Park and Tandou Property in relation to the total standing crop of potentially edible plants [Kinchega: reds $r = -0.80 + 1.14 (1 - e^{-0.007V})$, western greys: $r = -1.01 + 1.43 (1 - e^{-0.007V})$; Tandou: reds $r = -0.73 + 1.17(1 - e^{-0.007V})$, western greys $r = -0.64 + 1.30 (1 - e^{-0.007V})$]. Open symbols represent data not used in calculating the functions.

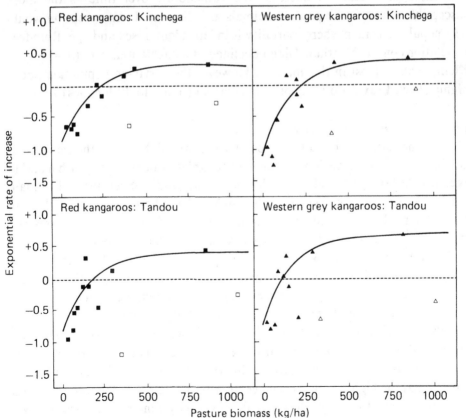

and grey (eastern and western greys combined) kangaroos across the entire west of New South Wales in 1984 (J. Caughley, pers. comm.). This anomaly was also reflected in the numerical responses plotted as a function of lagged rainfall, but the discrepancy was not as marked. An explanation may lie in different time lags between mortality and fecundity responses. Mortality is instantaneous and so is its detection by aerial survey. On the other hand the average detection lag for recruitment would vary depending on the reproductive stage of the females. Kangaroos born just after a drought would not be detected until 12-15 months later (the time it takes for birth, development within the pouch, and independence from the mother). The detection time lag is less for kangaroos born in times of plenty (J. Caughley *et al.*, 1984; Bayliss, 1985a) because most females are carrying advanced pouch young (up to six months old). Hence the post-drought recruitment surge in kangaroos would not be detected until 12-15 months later when young leave the pouch.

The abstract term 'maximum rate of decrease' is not particularly illuminating. The rate of decrease may depend on the current age distribution and sex ratio, varying because mortality is differentially distributed between age and sex classes. Rate of decrease may ameliorate over time as the more susceptible young, weak and old animals are eliminated. An exception would be a population crash where mortality is independent of sex and age. Reindeer populations on St. Matthew Island declined at a substantial rate ($a = -3.19$, Klein, 1968) almost in this manner. However the surviving population had a highly skewed sex ratio in favour of females (1 male to 40 females).

The maximum rate of increase

Most of the estimates of r_m for all populations are higher than the theoretical maximum rate obtained from modelled life tables of populations with a stable age distribution and under the highest conceivable conditions of female survival and fecundity. They are also higher than the maximum yearly rate predicted by body weight ($r = 0.40$, Caughley & Krebs, 1983) and the maximum rate obtained for both species in western New South Wales ($r = 0.33$ and 0.26 for red and western greys respectively, J. Caughley *et al.*, 1984). However, the individual point estimates used to calculate r_m in this analysis are each subject to considerable sampling error. Nonetheless such high rates are possible under the following conditions: a superabundance of food and an unstable age distribution and/or an imbalanced sex ratio in favour of females. Such conditions are not temporary aberrations but part and parcel of the boom and bust of juvenile recruitment in arid environments. Newsome (1977) showed that the highly unbalanced age distribution of red kangaroos in central Australia during the early 1960s was dominated by cohorts produced by good rainfall in the late 1940s, 1953-55, and 1958. He found also that sex ratio during a drought favoured females, particularly among animals more than three years old. It is highly likely that the post-drought (late 1960s) populations of the present study had both unstable age distributions and unbalanced sex ratios in favour of females. Johnson & Bayliss (1981) showed that even under average rainfall conditions in 1979 and 1980, red kangaroos in the study area had an unbalanced sex ratio in favour of females (0.7:1). Hence the term 'maximum rate of increase', as conventionally applied to the female segment of a population with a stable age distribution (Andrewartha & Birch, 1954), is an inappropriate concept for arid zone animals because it is not the maximum rate.

8.54 *Second-order effects*
The numerical response functions developed in this chapter assume that the main determinant of rate of increase of kangaroos is the level of food supply. Other factors undoubtedly affect the number of kangaroos, but are assumed to be of lesser importance. These include vegetational complexity and succession, time lags, age distribution, seasonality, year to year variation in weather, dispersal, territoriality, and spatial heterogeneity.

Caughley (1982) dismisses vegetational complexity because the dynamics of a one-herbivore-many-plants system can often be summarised by a one-herbivore-one-plant model. Dispersal, territoriality, diseases and parasites have been dismissed as unimportant in this chapter. Time lags have been accounted for in the models. Chapter 5 suggests that there is no seasonality in the proportion of plant groups (forbs, chenopods and grasses) eaten by kangaroos and that most of the differences in plants eaten is accounted for by extreme variations in food availability between consecutive three-monthly sampling intervals within a year. This leaves the previously discussed effect of age distribution on r_m. In terms of the whole grazing system, age distributions introduce a reactive lag which has a destabilising effect. This is an important influence on grazing system stability if the instability of the age distribution results from surges in recruitment (Caughley, 1981).

8.6 Conclusions
The annual rates of increase of red and western grey kangaroos on Kinchega National Park and an adjoining sheep station were plotted against indices of food availability. Rates of increase for ten years were plotted against annual rainfall lagged by six months. Rates of increase for four years were plotted against pasture biomass with no time lag. Rates of increase asymptoted at high levels of pasture biomass and when rainfall was above average. Rate

of increase was negative below 217 mm and 198 mm of rainfall per annum for western grey and red kangaroos respectively and at pasture biomass (mean of both locations) of 111 kg/ha for greys and 157 kg/ha for reds. No large differences were detected in the density trends of the two species. If one increased, the other increased. Similarly, the dynamics of kangaroos did not differ except in detail between the National Park and the sheep station despite the former having at least twice the density of kangaroos as the latter. The difference in density is attributed to sheep reducing the amount of food available to the kangaroos. Kangaroo numbers declined by about 50% during a drought, most dying over a single summer.

9

Condition and recruitment of kangaroos

NEIL SHEPHERD

9.1   Introduction

In this chapter body condition, nutritional intake, and recruitment of red and western grey kangaroos are related to trends in weather and vegetation. Some data relating to condition and recruitment of sheep are included for comparison. Body condition is also assessed as a wildlife management tool, with particular emphasis on assessment of kangaroo populations.

The kangaroo data are part of a long-term investigation that will not terminate until February 1986. Findings discussed here are mostly drawn from data collected between mid-1982 and the end of 1984. Methods will be published in detail elsewhere, but the essential information is provided below.

From June 1982 until December 1984 female kangaroos were sampled every three months on Kinchega and Tandou. Samples comprised 20 red kangaroos and 20 western grey kangaroos from Kinchega, and 20 red kangaroos from Tandou. An additional sample of 20 red kangaroos was taken on Kinchega midway between these three monthly samples; on one occasion (January 1983) this was extended to include 20 western grey kangaroos. These additional samples of red kangaroos from Kinchega were discontinued after February 1984 as was all sampling on Tandou. There are thus substantially more data on Kinchega reds than on either Tandou reds or Kinchega western greys. Western greys were not sampled on Tandou because they were at too low a density for samples to be readily obtained.

Kangaroos were found by spotlight from a vehicle and shot in the head with a high-powered rifle. Dissection was completed within 40 minutes of shooting. Sampling was spread across the study area.

Pastoral conditions from mid-1982 until December 1984 may be divided into crude categories on the basis of standing pasture biomass (Chapter 4). Categories used in this chapter are: 'bad' (samples 2-7; July 1982 − March

1983), 'transitional' (samples 1, 8, and 9; June 1982, April and June 1983), and 'good' (samples 10-18; July 1983 — December 1984).

9.2   Body condition

Several methods were used to estimate body condition: kidney fat index (KFI), body weight, bone marrow fat, blood counts, total serum protein, and muscle-bone ratio. Of these, KFI showed the most response to environmental change and most of the discussion will focus on it.

The KFI technique was developed by Riney (1955) and has since been used widely (Hanks, 1981). It was adapted to kangaroos by Caughley (1962). The technique used in this study differed from his only in that the left kidney was utilised instead of the right. A transverse cut was made through the perirenal fat at the caudal pole of the left kidney. The kidney and associated fat were removed, put in a plastic bag, and placed on ice for weighing in the laboratory. To obtain the KFI, perirenal fat weight was divided by kidney weight and multiplied by 100.

9.21   *Factors affecting variability of Kidney Fat Index*

Several factors affect either the variability of KFI or its usefulness as an indicator of trends in body condition (Hanks, 1981; Van Vuren & Coblentz, 1985). These were investigated in detail for kangaroos but, being outside the scope of this book, will be reported elsewhere. In summary, however, three points should be noted. First, when KFI is used to compare condition across seasons the sample should be restricted to females old enough to show a full fat deposition response under good pastoral conditions. In the kangaroos this corresponded to a crus length (approximates tibial length) greater than 450 mm. Second, given the observed skewness of KFI within samples, any statistical testing may require an examination of the underlying error structure. Third, within-sample variation induced by the effect of size of pouch young on KFI (a lactation stress effect) can be tolerated in the knowledge that major shifts in KFI will still be detectable between samples. If it was desired to fine-tune a continuing monitoring program on body condition a model to compensate for the effect of size of pouch young would need to be devised.

9.22   *Relationship between Kidney Fat Index and environmental variables*

This section explores the relationship between body condition (as measured by KFI) and three environmental variables: rainfall with various lags, pasture biomass (from Chapter 4), and quality of diet. Rainfall records used were the weekly totals from the Kinchega Ranger Station and therefore differ slightly from those used in Chapter 3. Dietary quality was measured in terms of nitrogen (modified Kjeldahl) and acid-detergent fibre (ADF) (modified Van Soest & Wine, 1968; and Van Soest, 1973). These were calculated as a percentage of dry weight of ingesta from the forestomach.

Since the peak levels of pasture biomass of the post-drought period were probably above the threshold producing maximum deposition of kidney fat, values for rainfall and vegetation biomass were transformed to natural logarithms to give a more realistic weight to these post-drought values. Four variables (rainfall, pasture biomass, % nitrogen, and ADF) were correlated with KFI with various lags. A lag term improved the correlation with KFI only for % nitrogen and ADF (6 weeks). KFI values were transformed to log values because of skewness, the resultant values being normally distributed.

Correlation coefficients and a summary of the least squares models are given in Table 9.1 and the untransformed data are graphed in Fig. 9.1. These correlation coefficients are not partial correlations (i.e. they have not been corrected for the effect of time) but are considered adequate to demonstrate the nature of the relationships. The correlation matrices (Table 9.1) indicate that each of the environmental variables is correlated strongly with KFI in both species with pasture biomass having the strongest correlation. Equally important, the environmental variables are tightly cross-correlated, fibre the least. Thus pasture biomass directly indicated the abundance of food and, indirectly through its correlation with nitrogen content of stomach contents ($r = 0.70$ for red kangaroos; $r = 0.88$ for western greys), the quality of that food. These results are therefore not at variance with the conclusion that green vegetation (an indication of high nitrogen content) is favoured by red kangaroos (Newsome, 1965a) and several other herbivores (Sinclair, 1975).

The relationship between pasture biomass and KFI appears different for reds and western greys (Table 9.1). Chapters 4 and 5 suggest that pasture biomass does not cover the range of vegetation eaten by either species of kangaroo. The difference in their dietary preferences (Chapter 5) therefore suggests that pasture biomass might not cover their food intake equally. It is not known whether the difference in pasture biomass-KFI relationships can be attributed simply to dietary preference, or whether habitat preferences (Chapter 7) restrict access of one or other species to components of the pasture.

Correlations between KFI and rainfall accumulated back from time of sampling for up to 56 weeks previously indicated a sharp increase in correlation when the period was extended to between four and eight weeks, and a subsequent gradual rise to maximum values as it was extended progressively

Fig. 9.1. Graphs of rainfall (6-weekly totals) and mean pasture biomass (measured 3-monthly), mean KFI (± SE) and its standard deviation, mean % nitrogen (± SE), and mean ADF (± SE) over the period of the study. Kangaroos are females from Kinchega National Park with crus >450 mm. Some standard errors are obscured by the data points.

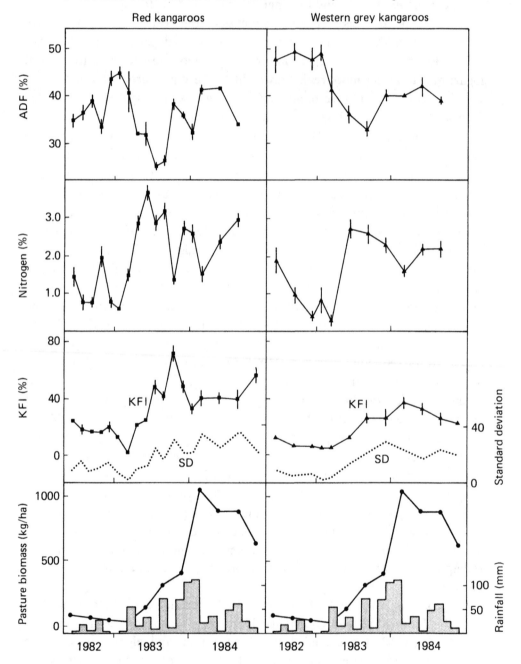

to 20 weeks (Kinchega reds, $r = 0.79$, $P\langle0.001$, % variance accounted for = 51.0) and 28 weeks (western greys, $r = 0.98$, $P\langle0.001$, % variance accounted for = 84.5). The correlation declined as the period was extended further. Thus the lag between rainfall and the accumulation of fat is around six weeks.

9.23   *Relationship between Kidney Fat Index and population parameters*

The previous section suggests that condition of kangaroos between June 1982 and December 1984 was primarily a function of nutrition. Over this period the amount and quality of vegetation fluctuated dramatically (Fig. 9.1). Given the tight relationship between pasture biomass and rainfall (Chapter 4) it is reasonable to assume that nutrition over this period depended

Table 9.1. *Correlation matrices by species for KFI, pasture biomass, total rainfall for 12 weeks preceding each sample, acid-detergent fibre (ADF)[1] and % nitrogen[1]. A summary of the multiple regression results is given for each species.*

		1	2	3	4	5
Red kangaroo [2]						
log KFI	1	1.00				
log biomass	2	0.87	1.00			
log rainfall (previous 12 wks)	3	0.78	0.89	1.00		
ADF lagged 6 wks	4	−0.66	−0.51	−0.51	1.00	
% nitrogen lagged 6 wks	5	0.73	0.70	0.69	−0.77	1.00
Western grey kangaroo [3]						
log KFI	1	1.00				
log biomass	2	0.97	1.00			
log rainfall (previous 12 wks)	3	0.93	0.93	1.00		
ADF lagged 6 wks	4	−0.85	−0.85	−0.86	1.00	
% nitrogen lagged 6 wks	5	0.86	0.88	0.91	−0.82	1.00

[1] Lagged by 6 weeks – the ADF and % nitrogen values now correspond to values for KFI etc. taken 6 weeks later.

[2] df = 15; 5% significance level for correlation coefficient = 0.48.
Regression model fitted:
    log KFI = 2.97 + 0.33 log biomass – 0.04 ADF lagged 6 weeks
    Percentage variance accounted for = 81.0

[3] df = 9; 5% significance level for correlation coefficient = 0.58.
Regression model fitted:
    log KFI = 0.35 + 0.42 log biomass
    Percentage variance accounted for = 93.8.

more on weather than on an interaction between weather, vegetation, and mammalian herbivore density; a conclusion also reached by Newsome (1965a) in his study of red kangaroos in central Australia. This is to be expected: unless managed (e.g. sheep), mammalian herbivore density cannot change as quickly in the arid zone as can the biomass of pasture.

This suggests that for arid-zone kangaroos a direct relationship should be sought between a measure of food availability and important population parameters such as adult survival, juvenile survival, and fecundity. However, kangaroos do not live by grass alone — other factors also influence these parameters and body condition is probably a reasonable summary of the impact of all factors on the population. The relationship between KFI and adult survival, juvenile survival, and fecundity will therefore be examined; adult survival here, and the other two parameters in the section on reproduction.

KFI over time is plotted for Kinchega reds and western greys in Fig. 9.1. The KFI of Tandou reds did not differ significantly from Kinchega reds on any sampling occasion. Correlations between KFI and a size-compensated index of bodyweight (bodyweight divided by crus) were examined for Kinchega reds and western greys. For reds the correlation coefficient was 0.40 (df = 306, $P\langle 0.001$) and for western greys it was 0.29 (df = 152, $P\langle 0.001$). These results suggest that KFI reflects real changes in well-being of the animals.

Drought mortality of red and western grey kangaroos was put at 41% and 62% respectively (Chapter 8). The finding of dead and moribund kangaroos during sampling associated with this part of the study (shooting adult females and ground counts for juvenile-adult ratios) suggested that the major mortality for western greys commenced at the end of November 1982 and for reds in late January 1983. Some animals of both species died before the respective major mortalities commenced. Relating these mortality patterns to changes in KFI we see two main things happening.

First, western greys had a significantly lower mean KFI than reds from June 1982 to January 1983 (Fig. 9.1). They also showed a more uniform decline in KFI, but managed to sustain a very low mean KFI for 8-12 weeks before the major mortality commenced. KFI of reds, in contrast, declined continuously over the period of high mortality. At the time of the respective major mortalities shot samples of Kinchega reds had a mean KFI of 6.5 ± 0.55 (SD) and western greys a mean of 4.6 ± 0.30 (SD). The remnant 'perirenal fat' in these animals was primarily the kidney capsule and remains of the ureters and associated connective tissue. For western greys this mean value corresponds closely with that of a group of starving kangaroos found stranded on small island in Lake Menindee in mid-1982 (mean KFI = 3.3).

Second, variance of KFI declined sharply for each species around the time of their major die-off (an impression can be obtained from Fig. 9.1). For some six weeks prior to this in Kinchega reds, and three months in western greys, the KFI distribution was markedly positively skewed (i.e. most animals were in poor condition but a small proportion were in reasonable shape). This is consistent with findings of de Calesta *et al.* (1977) who found experimentally that starvation mortality in mule deer (*Odocoileus hemionus*) was not completely synchronised. These authors hypothesised that individuals were at different condition levels at the start of nutritional stress and also had different abilities to maintain condition. Early deaths could therefore occur before numbers of moribund animals signalled an impending major mortality.

Low mean KFI coupled with a marked positive skew and small variance signalled impending disaster for red kangaroos. The KFI estimate was not sensitive enough to pinpoint the onset of mortality in western grey kangaroos, although a very small sample variance provided a clue. Measures of condition that moved later than KFI (e.g. muscle-bone ratio and bone-marrow fat) adequately monitored the critical period for this species.

Although mean KFI did not differ significantly between Kinchega reds and Tandou reds for any sampling period between June and December 1982, the mean KFI of Tandou reds was always above that of the Kinchega reds. Mean KFI of Tandou reds was low enough in March 1983 for a starvation-induced mortality to have commenced, but because there was no sample on Tandou in January, we cannot be sure when. Tandou reds may have been 'maintained' for longer than Kinchega reds by one or more of the practices adopted by the property management, including heavy shooting of kangaroos, removal of some sheep, and hand feeding of remaining stock.

Bodyweight change for animals in bad seasonal conditions was estimated for Kinchega reds and western greys by comparing their actual bodyweights with the bodyweight predicted by their crus length under good pastoral conditions. Table 9.2 summarises these data. The ability of around half the individuals to maintain bodyweight at the commencement of the major mortality is seen in western greys in December 1982 and in Kinchega reds in January 1983. However, by March 1983 very few animals of either species were maintaining bodyweight and the mean bodyweight depression of the lower quartile was similar to the percentage weight loss reported for mule deer that died of starvation (de Calesta *et al.*, 1975) and to two moribund western grey females from the Lake Menindee group mentioned earlier (bodyweight depressions of 29.7% and 26.3% respectively).

The logical conclusion from the KFI and bodyweight data is that deaths during the drought were due primarily to starvation. This is reinforced by absence of gross post mortem findings, other than excess peritoneal fluid, in almost all animals sampled.

The importance of these data is first, they confirm condition can be an important influence on, or indicator of, mortality (Sinclair & Duncan, 1972; Hanks *et al.*, 1976); second, they describe the relationship between one estimate of condition and mortality in kangaroos; and third, they suggest that no further nutritionally induced mortality of adult females occurred between March 1983 and the end of 1984.

9.3   Reproduction

Detailed reviews of kangaroo reproduction can be found in Tyndale-Biscoe (1973) and Russell (1974). The important features for red and western grey kangaroos are outlined below.

Red kangaroos have an oestrus cycle of 35 days. The gestation period is 33 days and a single young is produced. There is usually a post-partum oestrus and the embryo from this mating arrests development at the 80-cell stage until the previous young (now in the pouch) either dies or becomes ready to vacate the pouch permanently. The arrest mechanism is believed to be lactational inhibition, and under average seasonal conditions development recommences around 30 days before the current pouch young is due to depart. The period of pouch life is around 235 days, but the young suckles for a further four months from outside the pouch. The sex ratio of pouch young has usually been found not to depart significantly from parity, although two studies have shown a significant male bias (Newsome, 1965b; Denny, M. J. unpublished data, cited in Johnson and Jarman, 1983).

Most females in captivity mature sexually at 15-20 months post-birth

Table 9.2. *Summary of percentage deviations from the expected body weight under good pastoral conditions for kangaroos taken at selected sampling periods during the drought. Negative deviations are indicated (−). Zeros are included with positives.*

| | Western grey kangaroos | | | Red kangaroos (Kinchega) | |
	Dec 82	Jan 83	Mar 83	Jan 83	Mar 83
no. positive	8	5	0	7	1
no. negative	12	15	20	13	19
mean dev.(%)	− 6.9	− 7.9	−17.7	− 3.3	−14.9
mean dev. lower quartile (%)	−23.2	−18.8	−31.0	−11.2	−24.4

(Sharman & Calaby, 1964). In the wild these figures can be extended by up to a year depending on seasonal conditions (Frith & Sharman, 1964). Females can produce young until at least 12-15 years of age. Birth-rate is uniform throughout the year under average to good seasonal conditions (Caughley, 1962; Frith & Sharman, 1964; Sadleir, 1965) and the birth interval is around eight months. However, under adverse conditions advanced pouch young die and are continually replaced until seasonal conditions force the female to enter anoestrus (Newsome, 1964, 1965b).

There has been no published field study on reproduction in western grey kangaroos except for the studies by Pilton (1961) and Poole (1973). The former involved a small sample derived from widespread locations over several years and the latter failed to distinguish between eastern and western greys. Poole (1975) found the oestrus cycle to be 35 days and gestation 30.5 days in captive western greys. He recorded births throughout the year, but found a peak of births over summer and autumn (November-April inclusive) that accounted for some 75% of births. Breeding is thus described as seasonal in this species. Delayed embryos have not been recorded. Loss of a pouch young in the breeding season is usually followed by a return to oestrus after several days (Poole & Catling, 1974). Pouch life lasts for around 310 days in *M. fuliginosus melanops* (eastern mainland sub-species) (Poole, 1975) and the young at foot may suckle for a further six months. The interval between births in lactating western greys is around 12 months. Females usually mature sexually at 18 months of age, but a mean of 24.6 months to first offspring was recorded by Poole & Catling (1974).

9.31   *Reproductive strategies*

The standard parameters of reproduction in large mammals are fecundity by age and the adult sex ratio. Fecundity is measured as the number of live births (or female live births) per hundred females. It is influenced by season of births, frequency of births, and sex ratio of offspring.

Red Kangaroos

Red kangaroos are described in the literature as continuous breeders. Demonstrating this poses sampling problems — ideally a large sample of females is required each month to determine the pregnancy rate (nominally 8.4%). Use of smaller samples collected at longer intervals, or single large samples, requires ageing of pouch young to determine their month of birth.

Ageing confounds birth rates and pouch young survival. Caughley (1962) and Frith & Sharman (1964) attempted to avoid this problem by restricting the age of pouch young contributing to their estimate of birth rate below that at which substantial nutritionally-induced mortality would have commenced. Newsome (1965b) found that pouch young born in a severe drought survived 60.1 ± 7.6 days (SD). Use of pouch young less than 50 days old therefore appears adequate for most conditions.

The problem can be illustrated by comparing the distribution of births of Kinchega reds estimated by ageing pouch young up to 231 days (Fig. 9.2) with one based on pouch young ⟨50 days (not shown). The 231-day distribution indicates a much lower birth rate for June-December 1982. This discrepancy arises because pouch young survive for short periods during drought and there is a rapid turnover of these young until the female finally enters anoestrus. For any given month during June to December 1982 (the phase of increasing severity of drought) the 231 day distribution will utilise data from much further into the future than the ⟨50 day distribution. It will therefore contain more females that have finally entered anoestrus and will thus indicate a lower birth rate.

Interestingly, records of pregnant females in this study showed that in November and December 1982, 15% and 20% respectively of mature females would have given birth. These births were not even apparent in the ⟨50 day distribution, and therefore survival of pouch young was probably considerably less than 50 days at the height of the drought.

The proportion of anoestrus females on each sampling occasion can be calculated from Table 9.3 for red kangaroos on Kinchega and Tandou. The proportion increased as the drought progressed (max. 85% on Kinchega, 41%

Fig. 9.2. Births per 100 females by month for mature red kangaroos from Kinchega National Park based on ageing of all pouch young (range 0-231 days). Actual numbers of females contributing to each month are shown for each column.

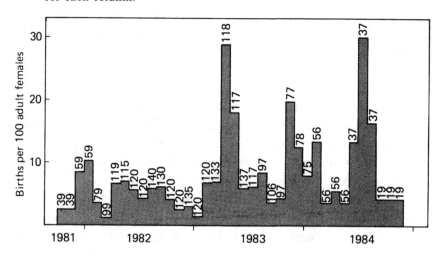

Tandou), but under good pastoral conditions only 2% of mature females were not reproductively active on Kinchega and all were active on Tandou. During the drought a small percentage of females was recorded as anoestrus whilst still carrying a pouch young (Table 9.3).

The April 1983 sample (first sample after breaking of the drought) is worth examining in detail. Of the 20 mature females collected on Kinchega, five had pouch young that were conceived before the drought broke (two born in 1982 and three in late February to mid-March 1983), two were anoestrus (both ⟨2 years of age, but classified as mature on skeletal size), and 13 (65%) had either just given birth or were in mid to late pregnancy. For this group of 13 the median birth date was 18 April and the median conception date was 16 March. Conception therefore occurred around two weeks after the rains.

The birth interval appeared to be eight months (Fig. 9.5). Only one set of twins was encountered and the maximum production rate of young per mature female was therefore 1.5 per annum. Sex ratios of pouch young are shown in Table 9.4. They did not depart significantly from parity.

Fecundity by age cannot be ascertained precisely from the data. However, the literature suggests the decline in fecundity after maturity can only be slight, and data from this study are consistent with that view. Age at first parturition under good conditions was around 16 months. Young animals that reached mature age in the drought were not reproductively active until after the drought broke (and from the April 1983 data possibly for some months after) and thus age at first reproduction became extended.

Adult sex ratios on Kinchega were derived from surveys of live animals in May 1984 and May 1985. In May 1984 the M:F ratio was approximately 1:1 ($n = 500$), and in May 1985 0.95:1 ($n = 817$). This is consistent with the ratio of 1.03:1 found during the mortality survey (G. Robertson, 1986, in press), but inconsistent with results of Johnson & Bayliss (1981), who reported sex ratios of 0.67:1 and 0.72:1 on large samples from surveys of the south-eastern part of Kinchega in 1979 and 1980.

Western Greys
Figure 9.3 shows births per hundred females derived from pouch young up to 301 days. This suffers the same theoretical disadvantages discussed above for reds. However, in this data set the survival problem is negligible, first because almost all pouch young died between the first and second samples (i.e. between June and September 1982), and second, because there was no attempt to replace them whilst pastoral conditions remained adverse.

A seasonal breeding pattern is clear from the post-drought years. In the

Table 9.3. *Summary of reproductive data from all samples. All mature[1] females are classified as pouch young + or − and within these classifications are categorised as pregnant (luteal (L)), oestrus (including pro-oestrus and post-oestrus (O)), quiescent (delayed implantation phase (Q)) or anoestrus (A). Numbers in categories are actual numbers of animals.*

	Date (month.year)	Pouch young +				Pouch young −				
		L	O	Q	A	L	O	Q	A	n
Red	6.82	2	2	14	0	0	0	0	1	19
kangaroo	7.82	0	0	15	1	1	1	0	2	20
	9.82	2	0	5	2	3	0	1	7	20
Kinchega	10.82	1	0	10	2	2	0	0	5	20
	12.82	1	0	4	2	4	0	0	9	20
	1.83	0	0	4	3	3	0	0	10	20
	3.83	1	0	1	0	0	1	0	17	20
	4.83	0	4	7	0	6	1	0	2	20
	6.83	0	2	15	0	2	1	0	0	20
	7.83	1	1	18	0	0	0	0	0	20
	9.83	0	0	15	0	3	1	0	0	19
	10.83	0	0	16	0	4	0	0	0	20
	12.83	8	0	8	0	2	2	0	0	20
	1.84	3	0	16	0	0	0	0	0	19
	3.84	2	0	16	0	1	0	0	0	19
	5.84	1	0	17	0	1	0	0	0	19
	9.84	0	0	13	0	5	0	0	0	18
	12.84	0	0	18	0	1	0	0	0	19
Red	6.82	1	0	15	1	2	0	0	0	19
kangaroo	9.82	2	0	4	4	3	0	0	0	13
	12.83	2	1	4	2	5	1	0	4	19
Tandou	3.83	0	0	2	0	2	6	0	7	17
	6.83	0	0	13	0	5	2	0	0	20
	9.83	1	1	17	0	1	0	0	0	20
	12.83	2	0	14	0	3	1	0	0	20
	3.84	0	1	18	0	1	0	0	0	20
Western grey	6.82	0	0	0	16	0	0	0	1	17
kangaroo	9.82	0	0	0	0	0	0	0	16	16
	12.82	0	0	0	0	1	1	0	17	19
	1.83	0	0	0	0	0	0	0	17	17
	3.83	0	0	0	1	0	0	0	16	17
	6.83	0	0	0	0	0	0	0	17	17
	9.83	0	0	0	1	1	0	0	14	16
	12.83	0	0	0	17	3	0	0	0	20
	3.84	0	0	0	18	0	0	0	1	19

Date	Pouch young +				Pouch young −				
(month.year)	L	O	Q	A	L	O	Q	A	*n*
5.84	1	0	0	18	0	0	0	1	20
9.84	1	0	0	13	1	2	0	0	17
12.84	0	0	0	17	3	0	0	0	20

[1] classification as mature based on skeletal size:
    reds = crus ⟩399 mm, forearm ⟩170 mm.
    western greys = crus ⟩415 mm, forearm ⟩165 mm.

1983-84 season 82% of mature females bred between September and December. This is earlier than described for captive western greys by Poole (1975), but consistent with data from Kinchega in 1977-78 (Shepherd, unpublished data) and the observations of Pilton (1961).

The almost total failure of breeding in the 1982-83 breeding season is consistent with the body condition of females at that time. More interesting is the absence of any reproductive response to the breaking of the drought. March and April are within the breeding season as described by Poole (1975)

Table 9.4. *Sex ratios of pouch young by species, location, and season expressed as males per hundred females.*

Species	Location	Pastoral conditions[1]	M/100 F	No.	Sig[2]
Red	Kinchega	Bad	93	54	ns
kangaroo		Good	131	148	ns
		Total	116	248	ns
	Tandou	Bad	81	38	ns
		Good	70	51	ns
		Total	89	102	ns
	Kinchega & Tandou	Total	107	350	ns
Western grey	Kinchega	Bad	125	18	ns
kangaroo		Good	83	88	ns
		Total	89	106	ns

[1] Sampling trips contributing to these categories are described in the text (Section 9.1). 'Total' rows include all sampling trips in which the species was collected at the location(s).
[2] Chi square tests, $P\langle 0.05$.

and some births were recorded in these months in 1984 (Fig. 9.3) and also in 1977-78 (Shepherd, unpublished data).

There was no evidence of delayed implantation (Table 9.3). The birth interval was around 11 months (Fig. 9.3) and this is consistent with Poole (1975). There were only two sets of twins and therefore the maximum production rate of offspring per mature female was 1.1 per annum. After September 1983 (the first post-drought breeding season), 98% of mature females sampled were classed as reproductively active (Table 9.3). The sex ratio of pouch young did not differ significantly from parity (Table 9.4).

Minimum age at first parturition was estimated to be around 18-19 months under good pastoral conditions. The seasonal birth pattern may extend this because most juveniles reach this age several months before the breeding season commences. Poor pastoral conditions will also delay first reproduction. There was no evidence of a decline in fecundity in older animals, but the sample size was too small to detect any slight downward shift.

The sex ratio of adults on Kinchega is not known. Data from Johnson & Bayliss (1981) provide two very different adult sex ratios: 0.55:1 ($n = 438$) in January-February 1979 and 1.03:1 ($n = 360$) in July 1980. It is suggested that accurate field sexing of western greys weighing less than 40 kg is difficult under conditions that are ideal for observation, and impossible under normal survey conditions. This precludes a reliable estimate of the ratio unless an unbiased sample of dead animals can be obtained. Robertson (1986) found a

Fig. 9.3. Births per 100 females by month for mature western grey kangaroos from Kinchega National Park based on ageing of all pouch young (range 0-301 days). Actual numbers of females contributing to each month are shown for each column.

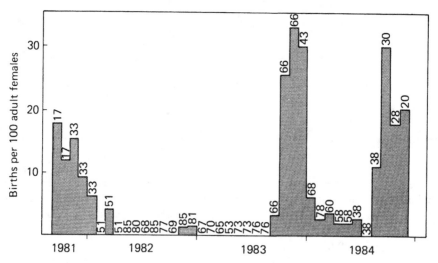

ratio of 1.1:1 in his sample of dead animals from the drought (n = 535), but any bias due to differential death rate between sexes is unknown.

9.32 *Survival of pouch young*

The literature suggests that birth rate in red kangaroos is uniform throughout the year under average to good seasonal conditions. The trend over time in mean pes length of pouch young (pes length = length of hind foot minus the claw) should therefore be a reasonable indicator of pouch young survival as pastoral conditions decline. Use of this indicator in western greys will be complicated by the seasonal birth pattern. Mean pes length is plotted over time in Fig. 9.4 for both Kinchega reds and western greys.

In section 9.23 it was suggested that the relationship between body condition of females and survival of pouch young should be examined. For Kinchega reds, correlations between log mean pes length of pouch young and log mean KFI of females from the same sample and from the previous sample were examined. Correlations between log mean pes length and the various explanatory variables that were correlated with KFI in section 9.22 were also examined. Results are given in Table 9.5. Whilst the correlations between pes length and KFI, and pes length and vegetation biomass, are highly significant, a substantial amount of the variance in mean pes length remains unexplained (30% with KFI). This may be due simply to the synchronised breeding of reds following the drought (Fig. 9.2) causing substantial changes in mean pes

Fig. 9.4. Mean pes length ($\pm$ SE) of pouch young from red and western grey kangaroos from Kinchega National Park over time. Pastoral conditions categorised in text (Section 9.1) are shown at the top of the figure. Open symbols are used where only one pouch young was found during a sampling period.

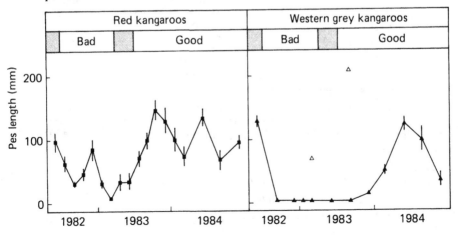

length in a cohort of pouch young (Fig. 9.4) independent of KFI changes (Fig. 9.1). Implicit in such an explanation is an assumption that a threshold value for KFI exists above which pouch-young survival is consistently high.

Correlation analysis was not performed on western greys because of the absence of pouch young over much of the sampling period. However, from their seasonal birth pattern (Fig. 9.3) large pouch young would be expected in late winter and early spring. This is apparent from the June 1982 sample, but there appears to have been almost total mortality of pouch young between June and September 1982. KFI was low in western grey females in June and around base level by September. Bone-marrow fat also declined sharply over this period.

Condition estimates of pouch young were also obtained. For samples of pouch young taken on Kinchega in good pastoral conditions the regressions of cube root of bodyweight on pes were calculated separately for reds and western greys. These regressions accounted for 96.9% of variance in reds and 97.3% in western greys. Bodyweight loss in poor seasons was calculated by:
Bodyweight loss $= 1 - (^3\sqrt{} \text{ actual bodyweight } / ^3\sqrt{} \text{ predicted bodyweight})$.
A summary of these data is given in Table 9.6. Large pouch young of both species were affected more severely than small pouch young. In reds there was a marked weight loss in the 100+ mm pes class after June 1982, accompanied by a rapid disappearance of this class from the population of pouch young (this class contributed 63% in June 1982 and only 11.5% in

Table 9.5. *Summary of correlation coefficients and the fitted regression model for mean pes length[1] of pouch young from Kinchega red kangaroos, with KFI and selected environmental variables.*

	Correlation coefficient	Sig.
log KFI	0.86	***
log KFI lagged 6 wks	0.71	**
log biomass	0.70	**
log rainfall 12 wks	0.64	**
% nitrogen lagged 6 wks	0.53	*
ADF lagged 6 wks	−0.46	n.s.

df = 14; significance levels for correlation coefficient; * $P\langle 0.05 = 0.50$; ** $P\langle 0.01 = 0.6$ *** $P\langle 0.001 = 0.74$
Regression model
    log mean pes = 0.973 log KFI + 0.884
    % variance accounted for = 70.6
[1] log mean pes weighted for variance with 1/(variance of log mean pes length).

pooled samples from July 1982 to March 1983). Newsome (1965b) also found that large pouch young were in poorer condition than were small pouch young in bad seasons.

It is worth noting that, despite the poor condition of the western grey females in June 1982, their large pouch young were on average only 4.5% below the bodyweight value they would have been expected to attain under good pastoral conditions. This suggests western grey females are prepared to maintain pouch young at the expense of their own condition. In contrast, red kangaroo females appear to maintain their own body condition at the expense of their larger pouch young.

9.33 *Proportion of juveniles*

The gap between pouch life and entry to the adult population was examined by assessing the proportion of juveniles in the population. The

Table 9.6. *Summary of condition assessment for pouch young during drought using regressions of $^3\sqrt{bodyweight}$ v. pes derived from young under good pastoral conditions. Numbers above and below the regression line are shown for small ($\langle 100$ mm pes) and large ($\rangle 100$ mm pes) pouch young in transitional (A) and bad (B) pastoral conditions. The numbers of females[1] contributing to these data and the percentage with pouch young are also shown.*

Pastoral conditions		Pouch young size pes (mm) $\langle 100$	$\rangle 100$	% Weight loss (mean) for $\rangle 100$ mm pes class	No. mature females	% Mature females with pouch young
Red kangaroo (Kinchega)						
A June 1982	No. above	3	1	− 5.9	19	95
	No. below	3	9			
B July 1982	No. above	20	1	−19.8	120	45
−Mar 1983	No. below	26	5			
C A + B	No. above	23	2			
	No. below	29	14			
Western grey kangaroo						
A June 1982	No. above	4	2	− 4.5	20	100
	No. below	4	10			

[1] maturity is based on skeletal size (see Table 9.3 for classification).
[2] $\langle 100$ mm pes class occupies around 54% of pouch life (Sharman *et al.*, 1964).

'proportion of juveniles' was the number of juveniles expressed as a percentage of the total number of kangaroos counted. A kangaroo was classed as a juvenile if less than two-thirds the size of an adult female. This class included animals up to 16 kg, about the size of first breeding in reds under good pastoral conditions.

Fixed transects along the road system were traversed by vehicle in the two hours after dawn. Speed was dictated by terrain, but around 50 kph was maintained unless adequate assessment of a group of kangaroos required examination through binoculars. The author was both observer and driver on all surveys. Four mornings were required to cover the transects and each transect was surveyed once on each sampling occasion. The proportion of juveniles was estimated each three months; for Kinchega from March 1982 to December 1984, and for Tandou from June 1982 to June 1983. As with other results for reds, results from Tandou mirrored those from Kinchega.

Results are graphed in Fig. 9.5 for Kinchega reds and western greys. Only the totals for each sampling occasion were used to calculate the percent juveniles. Total numbers observed varied for reds from 143 to 720 and for western greys from 37 to 337. The lowest totals for both species were from the March 1983 survey.

The decline in % juveniles of both species during the drought could have been due to several factors. First, there could have been a decreased adult mortality rate concurrent with a stable or increasing juvenile mortality rate. Second, an increased mortality of pouch young could have reduced recruitment into the juvenile population. Third, if recruitment was substantially reduced,

Fig. 9.5. Percentage of juveniles in the population shown at 3-monthly intervals for red and western grey kangaroos from Kinchega National Park. Pastoral conditions categorised in text (Section 9.1) are shown at the top of the figure.

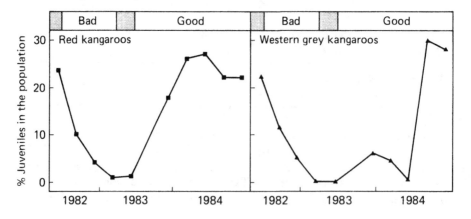

the decline could have been due to juveniles either dying or entering the adult population.

A substantial mortality of pouch young certainly occurred. Juveniles were also likely to have suffered a markedly increased mortality under the drought conditions that prevailed (Frith & Sharman, 1964; Newsome, 1965b). It is unlikely that many juveniles survived to enter the adult population.

Correlations between percent juveniles (arcsin transformed) and the same explanatory variables correlated with KFI in Section 9.22 were examined. A summary of the results is given in Table 9.7. For western greys all correlations were non-significant. For Kinchega reds the correlation between pasture biomass and percent juveniles was significant ($P\langle 0.01$), but left a substantial amount of variance unaccounted for (41%). These weak relationships are probably due in part to the scarcity of juveniles (particularly western greys) for much of the study period. However, any difference in the nature of the relationships between pasture biomass and pouch young survival and between pasture biomass and juvenile survival may also have an effect. Pouch young survival and juvenile survival are two major components of the percent juvenile estimate.

9.4 Performance of sheep

This section presents data on reproductive success and wool production of the Tandou sheep flock over the period 1980 to 1985 for comparison with trends in condition of the kangaroos. Reproductive success is assessed as the number of lambs marked per hundred ewes mated. Marking is the procedure whereby all lambs are tail-docked and earmarked, and males are

Table 9.7. *Summary of correlation coefficients for % juveniles with selected environmental variables. Data from Kinchega red kangaroos and western grey kangaroos.*

| | Correlation coefficients | | | |
	Red kangaroo	Sig.	W. grey kangaroo	Sig.
log biomass	0.77	**	0.35	n.s.
log rainfall 12 wks	0.55	n.s.	0.20	n.s.
% nitrogen lagged 6 wks	0.36	n.s	0.20	n.s.
ADF lagged 6 wks	0.00	n.s.	0.09	n.s.

df = 8; significance levels for correlation coefficients
* $P\langle 0.05 = 0.63$; ** $P\langle 0.01 = 0.76$.

castrated. On Tandou marking is carried out within six weeks of lambing; the marking percentage therefore confounds births with perinatal mortality. Lambing occurs in early winter (May-June). The data are presented in Table 9.8.

Reproductive success in sheep in the Australian rangelands is influenced by nutrition, flock husbandry, infertility diseases such as ovine brucellosis, and predation. Although pig predation had a substantial impact on marking percentage in the Tandou flock in some years (in 1977 a combination of pigs and a dry season lowered the marking percentage to 19%; R. Lukoschek, pers. comm.), trends in 1980-85 generally followed the vegetation trends described in Chapter 4. This suggests that nutrition was the major factor determining reproductive success over the study period.

Although annual stock numbers were maintained by purchases early in 1983, the drought saw the loss of 5,500 sheep comprised of 2,500 older ewes and 3,000 marked lambs. Some of these losses were attributed to diseases associated with commencement of grain feeding in mid-1982, but without this grain feeding total losses by March 1983 would undoubtedly have been much higher. The 1983 lamb marking percentage undoubtedly also reflects the benefits conferred on the sheep by supplementing the meagre amount of pasture available.

The wool production figures of Table 9.8 estimate adult fleece weight as total woolclip (kg) divided by the number of adult sheep equivalents. Adult sheep equivalents were derived by a complex process that involved estimating the contribution of the previous year's lambs to the shearing tally (including allowance for post-marking losses), averaging the fleece weights of ewe and wether hoggets at 3.5 kg, calculating the value of their contribution in terms

Table 9.8. *Lamb marking percentage and average fleece weight from 1980 to 1985 for merino sheep on Tandou. Derivation of the average fleece weight is described in the text.*

Year	Lamb marking percentage	Adult sheep equivalents	Wool clip (kg)	Av. fleece weight (kg)
1980	63	13637	77515	5.68
1981	72	10702	63188	5.90
1982	44	12388	62258	5.02
1983	57	13837	58460	4.22
1984	92	8741	61167	6.99
1985	80	9903	69095	6.97

of adult sheep, and then adding this number to the number of adult sheep actually shorn. The resultant average fleece weights are at best a rough approximation, but results derived by this method correspond closely with results from fleece weight measurements made as part of the 1985 shearing (6.97 v. 6.8).

Sheep were shorn in January until 1982 and in March-April thereafter. Stock were bought and sold during most years, so the pastoral conditions on Tandou were not the sole influence on fleece weight or sheep numbers. Despite these problems, the trend in estimated fleece weights corresponds well with the trends in rainfall and pasture biomass described in Chapter 4. A similar effect of adverse pastoral conditions on fleece weight was demonstrated by Brown *et al.* (1968) for an experimental flock at Cunnamulla in Queensland, and Robards (1978) presented data showing a within-year effect of seasons on wool growth that he believed was nutritionally based.

Trends in reproduction and condition (fleece weight) of the sheep are similar in kind to those presented for kangaroos. The magnitude of the changes is obviously different. In part this is due to fleece weight having a much smaller response range than KFI (fleece weight rarely varies more than 30% between good and bad seasons). However, sheep management practices over the period, particularly supplementary feeding, probably also contributed substantially.

9.5 Discussion

The best measure of a herbivore's response to pastoral conditions is its rate of increase. Air survey is commonly used to obtain the necessary indices of abundance, but aerial survey of kangaroos has its problems: differential visibility of different species in different habitats, biases induced by environmental variables, and some insensitivity caused by a time lag between birth and sightability from the air (Short & Bayliss, 1985, and Chapter 8). Hence information is required to supplement rate of increase calculated from aerial survey estimates. Obvious candidates are factors affecting rate of increase such as age structure (Wilson, 1975), recruitment, and body condition (this chapter). The latter has been used extensively on ungulates in New Zealand, Africa, and North America (see Hanks, 1981, and studies reviewed therein), but has not been used in Australia for kangaroo management.

There have been few determined efforts to link body condition with rate of increase or with two of its components, birth rate and death rate. Components of rate of increase tell us most things we need to know about a population for management purposes and provide a yardstick by which to measure change.

Body condition may track that change (or anticipate it), but estimates of body condition are useful only if they can be related to rate of increase or to its important components.

Both body condition and rate of increase are point-in-time estimates summarising what has already happened to the population. Changes in body condition usually precede major changes in rate of increase. This is illustrated by the St. Matthew Island reindeer mortality (Klein, 1968). Klein noted a downward shift in components of rate of increase (particularly recruitment) between 1957 and 1963, but in 1963 the index would still have been positive. However, between 1957 and 1963 there was a marked downward shift in body condition indicated by decreases in mean bodyweight and body size. Hanks (1981) used these data to criticise rate of increase as a predictive tool for wildlife management and to demonstrate the usefulness of body condition estimates. However, the more appropriate lesson from the reindeer mortality is that neither body condition nor rate of increase tells the wildlife manager what will happen to a population. To use either of them predictively requires an understanding of the dynamics of the plant-herbivore system. Even then, their predictive value is only as good as the manager's ability to predict the weather, human imposts, or other environmental changes (see Croze *et al.*, 1981, with respect to elephants).

Although there is potential to quantitatively link body condition with reproduction, survival of pouch young, and survival of juveniles in kangaroos, body condition data provided only qualitative information on the chances of an adult kangaroo living or dying. The major reason was that the body condition estimators were at or near the bottom of their range at the time of the major mortalities and there was little scope for change that would allow for high correlation between the body condition estimator and a rapidly changing mortality rate. However, for most management purposes it would be sufficient for body condition measurement to indicate a potential for major mortality if nutrition were not to improve.

In the context of environmentally-induced change, body condition in kangaroos was correlated highly with food supply, but less so with rainfall, the major predictor of food supply. Reproduction and pouch-young survival were also more highly correlated with body condition than with environmental variables.

Data presented in this chapter raise interesting questions about the relative fitness of reds and western greys in an arid environment. Bayliss (1985a) found that within the variability of his air survey data there was little difference in rate of increase between red and western grey kangaroos over the period 1973-81 on Kinchega or the surrounding properties. However, data summarised in the next three paragraphs suggest that, over the period of this study at least, red kangaroos appeared to cope better than western greys.

Body condition measurements clearly favoured reds, both in this study and in 1977-78 (Shepherd, unpublished data). However, it should be noted that the main techniques used (KFI, bone-marrow fat, and muscle-bone ratio) showed different responses between species, apparently intrinsic rather than just a difference in timing. Until further work is done caution is required in comparing species on the basis of body condition.

Mortality of western greys occurred earlier and was more severe (Robertson, 1986, in press). This difference should not be over-interpreted in terms of fitness because, had the drought lasted another month, losses might have been more even. In March 1983 reds showed a continuing drop in KFI but the western greys showed an increase; losses in reds to that point were in the so-called vulnerable age groups (old and young) whereas losses in western greys were spread across all age classes. The worst of the mortality was probably over for western greys and any further losses would have been at a much reduced rate. This pattern was suggested for red kangaroos during an earlier drought (Frith, 1964).

It is in reproduction that the most obvious differences occur. Reds mature earlier and, in good conditions, have a much higher production rate per adult female. In bad conditions they are better equipped to take advantage of improvements than are western greys. The full significance of this latter point remains to be determined, but on data from this study up to six months could be lost between the end of a drought and commencement of breeding in western greys. Whether the delay is a function of season or body condition, or both, is unclear. The termination of the drought near the termination of the breeding season confounds these possibilities in this study. The problem could be addressed in a simple experiment on captive western greys by terminating nutritional stress at the beginning and middle of the breeding season.

If reds survive better (Chapter 8), reproduce faster, and utilise the habitat more broadly (Chapter 7) why is their rate of increase not significantly higher? The answer may lie in the rate of survival of pouch young and juveniles. Western greys have a longer pouch life and also suckle from outside the pouch for longer than reds. Female western greys also appear to support pouch young at greater cost to themselves than do reds. Whilst a strategy aimed at maximising the efficiency of reproduction was not successful for western greys in the 1982-83 drought, it may well have been during the run of good and average seasons covered by Bayliss (1985a).

The decline in reds suggested by aerial surveys of Kinchega between June

1983 and June 1984 (Chapter 8) are inconsistent with trend of body condition and reproduction reported here. The body condition data argue against nutritionally-induced mortality between March 1983 and December 1984 and the proportion of juveniles in the population over the same period suggested no major mortality of juveniles. Other indirect evidence also suggested that red kangaroos neither died nor moved in numbers sufficient to account for the reported decline, and in fact the reproduction data and juvenile counts suggest an increase should have occurred. Since the population estimates for western greys derived from these two surveys are entirely consistent with data on body condition, reproduction, and movements it is suggested that some factor(s) other than cloud cover and temperature (Chapter 8) affected the results for red kangaroos on one or both of these surveys.

These observations suggest that major management decisions for kangaroos should not be based solely on a single air survey result, particularly if it is inconsistent with other evidence concerning the population or environmental conditions. They in no way reflect on the trends obtained from a series of aerial surveys where unusual effects might reasonably be expected to even out over time.

9.6   Conclusions

Body condition of kangaroos is linked to some components of rate of increase, such as fecundity, pouch-young survival, and adult and juvenile survival. Overall, body condition appears a useful summary statistic of a kangaroo's response to its environment. Body condition is clearly a useful tool for researching relationships involved in the animal's response to its environment. Its value as a management tool is questionable if more direct measures of numerical response are available and there is a known relationship between these and readily measured environmental variables.

Body condition was correlated best with food supply. A correlation between food supply and rainfall has already been demonstrated in Chapter 4. This provides strong circumstantial evidence from the animal to suggest that the trends in rainfall and kangaroo abundance described in Chapter 8 and by Caughley *et al.* (1984) are causally linked.

Red and western grey kangaroos adopt quite different strategies for reproduction and maintenance of condition. Data from this chapter suggest that reds should have an edge in the Menindee environment, but the similar rates of increase reported in Chapter 8 suggest any such advantage is small.

Although interpretation of the sheep data requires considerable caution, results for both reproduction and condition provide support for the notion that the responses of sheep and kangaroos to pastoral conditions are very similar.

10

Ecological relationships

GRAEME CAUGHLEY

This chapter briefly summarises the findings presented in the previous chapters and attempts a synthesis.

10.1   Weather

The weather at the Menindee sites is a sample of that of the Australian sheep rangelands as a whole. Annual rainfall over 100 years has averaged 236 mm with a standard deviation among years of 107 mm, giving a coefficient of variation (CV) of 45%. Serial correlation of rainfall between successive years is very weak at $r = 0.13$.

Whereas daily temperature has a marked annual cycle it is not paralleled by a similar seasonality of rainfall. The serial correlation between rainfall in the same season across successive years was estimated as $r = 0.05$. Hence rainfall is unpredictable across a year and unpredictable from one year to the next. In three of ten years the annual rainfall is more than fifty percent above or below the annual average. Floods and droughts are common.

10.2   Plants

The high variability of rainfall leads to a much higher proportion of annuals in the pasture than on any other continent. The dominant perennials are low shrubs with deep roots. The pasture layer waxes and wanes according to rainfall. One suite of predominantly annual species germinates with rain in winter. Another suite responds to summer rain.

We saw in Chapter 4 that the pasture biomass could be predicted from rainfall over the previous six months. The relationship was largely independent of time of year and influenced only weakly by soil type. Even more predictable was the growth increment in the absence of grazing over any period of three months. Ninety-seven percent of variance was accounted for by a combination of rainfall over these three months and pasture biomass at the start of that interval. Of these the rainfall was the more important but starting biomass had a significant negative effect. Thus although the rainfall was unpredictable, the response of plants to any set of rainfall events was highly predictable. The swings of weather between wide extremes generate sympathetic swings in pasture biomass which may pass through two orders of magnitude over a year.

10.3   Herbivores

Of the three dominant herbivores in the region of study, red kangaroos breed throughout the year, western grey kangaroos are capable of breeding in all months of the year but there is a sharp peak of births between September and December (Chapter 9), and sheep breed seasonally with the birth pulse in this area arranged for May-June.

The dynamics of the two species of kangaroos were explored in Chapters 8 and 9. Plant biomass was identified as the proximate influence which was itself driven by rainfall. The dynamics of the sheep were artificial, changes in density being influenced more by human decisions than by the net effect of birth and death.

All three species changed diets in response to changed pasture, the red kangaroo the least (Chapter 5). All moved locally to take advantage of the changing pattern of the vegetational mosaic but none was naturally nomadic. Dispersal was not detected as an important influence on dynamics, most of the animals living and dying within relatively small home ranges (Chapter 7). The offtake of pasture per individual animal rose with increasing pasture biomass and levelled out at high biomass (Chapter 6).

10.4   Relationships

The picture sketched so far is of an environment with one driving variable — rainfall — which is unpredictable. However the plants react to a given episode of rain in a predictable manner and the herbivores' dynamic reaction to the change in vegetation is also quite predictable. The relationships within the system are tight. Only the input of water is capricious.

10.41   'Centripetality' of the system

The usual word is 'stability' but that has two meanings within ecology. The first, corresponding to its every-day use, 'implies temporal stability, and has to do with the amount of change, or rather lack of change, which a population or system exhibits over time' (Walker & Goodman, 1983). In this sense the Menindee system is unstable. The second usage refers to the speed with which a system returns to its equilibrium state after being displaced from it (Walker & Goodman, 1983) and is close to the sense intended here. However 'centripetality' rather than 'stability' is used to add the connotation that the forces causing temporal variation may be so powerful, continual and multidirectional that the 'equilibrium' is seldom or never occupied. The forces of centripetality are those that dampen temporal variation; a centripetal system is one that would come to equilibrium if it were not buffeted continually.

Centripetality implies feedback within the system's dynamics. Two loops have been identified. The first is the inhibiting effect of pasture biomass on pasture growth: the more biomass to begin with the less growth. The second is the effect of herbivores on pasture biomass: the more herbivores the more pasture eaten. How these two feed-back loops contribute to the centripetality of the system is examined in the next section.

10.42   *Strength of centripetality*

The contribution of feedback loops to the system's dynamics was examined by deducing the behaviour of the system in the absence of those loops. Sequences of hundred-year weather were generated with the same means and variances per three-month period as the observed weather at Menindee between 1884 and 1983. Successive increments to pasture biomass were estimated from simulated three-month rainfalls, offtake was estimated from the amount a kangaroo eats when confronted with a given biomass of pasture (the functional response: Chapter 6) and the change in kangaroo density was estimated from the relationship between rate of increase and pasture biomass (the numerical response: Chapter 8) to give the population growth increment of herbivores over the three months. The books were balanced weekly. These projections make no allowance for long-term changes in the structure of the system that are independent of the state of the system now, or that may be carried within the system as a memory of conditions long passed.

System A: Pasture with biomass feedback but without herbivores
The rainfall in millimetres each three months was a random draw from a normal distribution with the following observed means and standard deviations (Chapter 2):

	Mean (mm)	SD
Dec-Feb	62	59
Mar-May	57	47
Jun-Aug	59	34
Sep-Nov	61	44

The growth increment or dieback decrement to the pasture in kg/ha was calculated from those rainfalls as

$$\Delta V = -55.12 - 0.01535V - 0.00056V^2 + 3.946R$$

where ΔV is the increment over three months in the absence of grazing, V is the pasture biomass at the beginning of those three months and R is the rainfall in millimetres over that period (Chapter 4). The growth increment not accounted for by V and R provided a standard deviation around the regression of 52 kg/ha (see Section 4.32). The increment was taken as a random draw from a normal distribution with mean equal to the solution of the above regression equation and with a standard deviation of 52. It was added to pasture biomass and the iteration continued.

The long-term average of five independent runs of 100 years was 521 kg/ha of pasture with a standard deviation between years of 121 kg/ha. Variation of the level of pasture biomass initiating each run indicated that System A took less than ten years to forget its history.

System B: Pasture without biomass feedback or vertebrate herbivores
This system was the same as System A except that the equation for generating three months of growth (kg/ha) was contracted to

$$\Delta V = -55.12 + 3.946R$$

by dropping out the terms in V while retaining the standard deviation around the regression of 52 kg/ha. Pasture growth and dieback were then a linear function of rainfall, lacking the feedback loop that reduced growth as pasture biomass increased. This system was not centripetal. Pasture biomass, although fluctuating from year to year, increased progressively at a mean annual instantaneous rate of $r = 0.06$ (Table 10.1).

Alternatively, the growth increment equation for this system could be described by

$$\Delta V = -109.45 + 3.873R$$

which is the empirical regression of growth increment on three-month rainfall, the effect of starting biomass being ignored. This system (B1 of Table 10.1) behaves similarly to system B except that rate of increase of pasture biomass is reduced to $r = 0.03$ on an annual basis.

System C: Pasture with biomass feedback and grazed by kangaroos
System A was here extended by adding a trophic level. To simplify the system only red kangaroos were added. They average 35 kg in weight and over three months each eats $86(1 - e^{-V/34})$ kg (see Chapter 6).

Several numerical responses are given in Chapter 8, both in terms of rainfall and in terms of pasture biomass. Bayliss chose the Ivlev curve of the form

$$r = -a + c \left(1 - e^{-dV}\right)$$

where r is instantaneous rate of increase on a yearly basis, a is rate of decrease in the absence of food, $c - a = r_m$ is the kangaroos' intrinsic rate of increase, and d is the slope of the curve relating r to V. That formulation is followed here but $r_m = 0.4$ is chosen as the estimate of intrinsic rate of increase, being near the middle of Bayliss's range of estimates. Maximum rate of decrease would have been underestimated by Bayliss's linear extrapolation method so we use an instantaneous rate of $a = 1.6$ on a yearly basis to reflect the observed decline over four months during the 1982-83 drought. The estimate

Table 10.1. *Mean, standard deviation (s) and coefficient of variation (CV%) among years of simulated 100-year trajectories of pasture and kangaroos subjected to a mean annual rainfall of 236 mm and a standard deviation among years of 107 mm. Values are averages of five runs.*

Structure of system	Pasture			Kangaroos		
	kg/ha	s	CV%	n/ha	s	CV%
A. No kangaroos	521	121	23	—	—	—
B. No feedback between plant growth and biomass; no kangaroos	Pasture increases at $r = 0.06$ annual			—	—	—
B1. B modified; see text	Pasture increases at $r = 0.03$ annual					
C. Pasture biomass feedback and kangaroos	295	115	39	0.39	0.25	64
C1. C modified; see text	335	147	43	1.19	0.58	49
D. No pasture biomass feedback; kangaroos	625	567	91	1.26	0.90	71
E. Kangaroos present but no pasture offtake	521	124	24	Kangaroos increase at $r = 0.29$ annual		
F. As for C but no annual variation in rainfall; pasture growth stochastic	248	70	28	0.60	0.17	28
G. As for F with no annual variation in rainfall; pasture growth deterministic	230	22	10	0.68	0.09	13

of $d = 0.007$ is Bayliss's value. Thus the numerical response of red kangaroos to pasture biomass in System C becomes

$$r = -1.6 + 2(1 - e^{-0.007V})$$

By this formulation the kangaroos increased when pasture biomass exceeded 230 kg/ha and decreased when it fell below that level.

The equation linking plant growth and rainfall is as for System A but its slope is reduced from 3.946 to 2.5. That is to allow for the effect noted in several studies (Sharrow & Motazedian, 1983, and studies quoted therein) that our method of estimating offtake by the difference between caged and uncaged plots results is an overestimate. That conclusion is reinforced by the shape of the curves in Fig. 4.3. They lack the left limb of the parabola relating growth to biomass over a short period (see particularly the studies reviewed by Noy-Meir, 1975). Thus the original equation is adequate for System A where kangaroos are absent but must be modified to allow for their presence in System C. This reasoning conflicts with that advanced by McNaughton (1979) who showed that grazing increased the growth of pasture in the Serengeti. The contradiction need not be taken too seriously. McNaughton dealt with perennial grassland episodically grazed down from high biomass. We dealt with a largely annual pasture grazed continuously as it grew up from low biomass. If growth increment per day is approximately parabolic on pasture biomass, and the evidence for that is strong, the effect of grazing on growth will be negative below a threshold level marked by the position of the parabola's peak and positive above that threshold. The reduction in the slope parameter is largely arbitrary but its effect was to reduce long-term mean density of kangaroos to a plausible level. Simulated results for System C are in Table 10.1.

An alternative tactic is to hold the constant for the effect of rainfall at its measured 3.946. This modified system is coded C1 in Table 10.1. It differs from System C in having a much higher mean density of kangaroos whereas mean pasture biomass is increased only marginally. The modification has little effect on dynamics or centripetality, indicating that conclusions and interpretation are not dependent upon the absolute value (as against the sign) of that constant.

Figure 10.1 shows a single, but typical, 100-year run for System C. Yearly values for rainfall are annual totals. For pasture and kangaroos they are means of four seasons per year. The system fluctuated wildly but the kangaroos were not lost from it during this or any other run.

An examination of the troughs of pasture biomass in years 45 and 75 allows some feeling for the behaviour of this system. The first reflects die-back caused by low rainfall. The second, and deeper, trough is caused largely by heavy

Fig. 10.1. The bottom histogram shows modelled yearly rainfall drawn from the Menindee climate, the middle graph the pasture biomass that would be generated by that rainfall and reduced by the grazing of kangaroos, and the top graph shows the expected density of kangaroos whose rate of increase is a function of that pasture biomass.

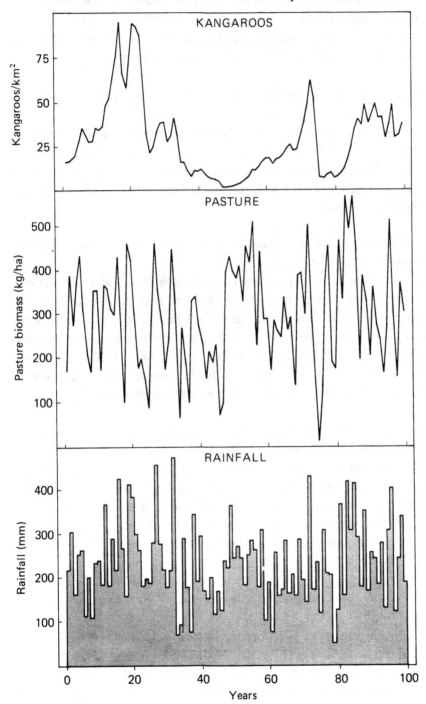

grazing. The lowest annual rainfall (year 79) produced another trough but this was not as deep because kangaroos were then at low density. Thus the annual ups and downs of pasture biomass are a product of the interaction between rainfall over the year and density of herbivores. The latter is determined not by current conditions but by the history of previous biomass levels which determine current density and hence current grazing pressure.

System D: Pasture without biomass feedback but with kangaroos eating it
Functional and numerical responses of kangaroos in System C are here married to the vegetation dynamics of System B which lacked a feedback between plant growth and plant biomass and which as a consequence lacked centripetality. The regression slope of plant growth on rainfall is 2.5 as in System C. Results in Table 10.1 indicate that addition of kangaroos restored the system's centripetality.

System E: Pasture with biomass feedback; kangaroos react dynamically to pasture biomass but do not affect its level
The absence of an effect of kangaroos on pasture is intended to simulate a distinct possibility for an arid-zone grazing system: that the dynamics of the vegetation are so dominated by rainfall or its lack that grazing is inconsequential to it, but yet rate of increase of kangaroos is determined by pasture biomass. The dynamics of this system lacked centripetality (Table 10.1), the kangaroos tending to increase indefinitely at an average annual rate of $r = 0.29$. The pasture component of the system retained centripetality.

System F: Pasture with biomass feedback; grazing by kangaroos; no year-to-year variation in rainfall; moderate stochasticity
This arid-zone system has the rate variables of System C without its variation in rainfall. It is driven by the mean rainfall of an arid-zone system but with the variance of temperate-zone rainfall. The variation still present is generated by the standard deviation of the pasture growth reflecting influences not accounted for by rainfall or pasture biomass. The results in Table 10.1 are for comparison with those of System C.

System G: Pasture with biomass feedback; grazing by kangaroos; totally deterministic
System G is system F stripped of its limited stochasticity. The results of Table 10.1 indicate the behaviour of a totally predictable system. The standard deviation between years reflects only the changes in kangaroo numbers and pasture biomass over the first ten years as the system moves to equilibrium.

10.43   *Comparison of systems*

These simulations are not total reality. They are simplified systems that lack the nuances of real systems. It is to be hoped however that their dynamic behaviour does not differ in kind from that of real systems. We have gone to some lengths to incorporate within them what we guess to be the dominant properties of real systems but that is no guarantee that the conclusions reached from these simulations are correct. Specifically, simplifications include dealing only with red kangaroos, denying them an age stucture, treating the species of the pasture, each with its unique ecology, as a homogeneous porridge of vegetation, and assuming no long-term trend in the ratios of plant species within the pasture. Equally, the simulations may have got relationships right but rate variables wrong. The obvious conclusion is that interpretation must be conservative. Absolute values of long-term densities and their year-to-year variation should not be taken too seriously. The ranking of these between systems is all we seek.

With those cautions recorded a few tentative conclusions may be offered:

(1) The depression of pasture growth by the feedback from pasture biomass is necessary, in the absence of grazing, to maintain the system's centripetality (System B compared with System A).

(2) The feedback between offtake of pasture by kangaroos and biomass of the pasture is sufficient to eliminate any long-term trend in pasture biomass even when there is no feedback from pasture biomass to pasture growth (System D).

(3) Were the effect of rainfall on vegetation so powerful that the observed swings in pasture biomass would be largely independent of grazing by kangaroos, the pasture component of the system would be centripetal but the kangaroo component would not (System E). This is no trivial conclusion. I was convinced at the beginning of the study that this model was the correct one for kangaroos and pasture, that the kangaroo populations were tossed around by swings of pasture biomass that they were powerless to modify significantly. In the light of these simulations that 'bottle-cork boat on the Irish Sea' model seems unlikely.

(4) The effect of kangaroos on the system is to reduce the long-term average of plant biomass but to increase relative fluctuation (as measured by coefficient of variation between years) around that mean (System A compared with System C).

(5) The year-to-year variation in weather produces a long-term mean of pasture density and kangaroo density differing from that in the absence of variation. The more the rainfall varies around its long-term average the higher is the long-term average pasture biomass and the lower is the long-term

average density of kangaroos (Systems C, F and G compared). That effect is a consequence of the non-linearity of the numerical response of kangaroos to rainfall: a shortfall of rain reduces the kangaroos' rate of increase more than an excess of the same magnitude increases it. Symmetrical deviations around mean annual rainfall have an asymmetrical effect on the dynamics of the kangaroo populations.

10.44   Dynamics of the system

Long-term dynamics

The previous section explored what was likely to happen to the system if various relationships were censored from it. That exercise indicated that the state of the system at any particular time is influenced strongly by the vagaries of rainfall, that the system fluctuates violently in response to these events which are essentially random in spacing and intensity, but that it is not entirely at the mercy of a table of random numbers. Without negative feedback from pasture biomass to pasture growth the system would break down in the absence of grazing. That conclusion is not exciting. It tells us only that if plants were capable of growing to indefinite size and density they would do so. That feedback loop is not needed to maintain centripetality if the pasture is grazed by kangaroos, but with grazing substituted for plant biomass feedback the force of centripetality (measured by coefficient of variation of plant biomass among years) is, according to the calculations summarised in Table 10.1, about four times weaker than that provided by the plant biomass feedback itself.

Grazing by kangaroos provides a second negative feedback loop taking the form: more kangaroos, more pasture eaten, less pasture available, fewer kangaroos. The calculations of Table 10.1 argue that in the absence of this loop the grazing system could not persist. It would be fractured by the kangaroos' continuing tendency to increase. Whereas the effect of that loop is intuitively obvious when a constant environment is considered, it is less obvious that the loop would confer centripetality on a system whose major swings of pasture biomass are virtually independent of kangaroos, and where those fluctuations have a mean wave length of the same order or less than the length of a generation of kangaroos. Nonetheless, the results of these calculations argue strongly that the grazing feedback is indispensable to the centripetality of this grazing system.

A third feedback loop is possible but unlikely for kangaroos: the influence of herbivore density *per se* on the rate of increase of those herbivores. Such has been detected for populations of small herbivorous mammals (Krebs, 1971) where it acts through spacing behaviour, with or without territoriality, generating a rate per head of dispersal that increases with density. Territoriality can be dismissed for kangaroos because their home ranges overlap (see Chapter 7). We did not find evidence of spacing behaviour working through alternative mechanisms and nor has any other study of kangaroos. But we do not have the data allowing a categorical rejection of all spacing behaviour as an influence on the dynamics of kangaroo populations. That experiment remains to be done.

To complete the possibilities, a feedback loop between kangaroo density and the density of their predators or prevalence of diseases could tighten the system's centripetality. That loop essentially is lacking at Kinchega National Park. Young kangaroos are a very minor component of the diet of foxes (*Vulpes vulpes*) and wedge-tailed eagles (*Aquila audax*). Dingos (*Canis familiaris*) are absent. Disease was not identified as an important agent of mortality although it was specifically searched for.

Hunting is common on sheep stations surrounding the park (see Fig. 8.3). If the *proportion* of kangaroos shot increased with their density then centripetality would be tightened. If the *number* shot increased with density but the proportion remained constant, centripetality would be little affected. If the proportion shot increased as density decreased the loop would provide positive feedback that would weaken centripetality. Hunting does not necessarily reduce the coefficient of variation of kangaroo density among years. There is abundant evidence, of which Fig. 8.3 provides but one example, that the proportion of kangaroos shot each year in the sheep rangelands tends to increase during droughts, which is when the populations are declining (Caughley, Grigg & Smith, 1985). There is no evidence that the proportion shot tends to rise as density increases. Hence hunting of kangaroos seems an unlikely enforcer of centripetality and may in some circumstances weaken it.

The centripetality of pasture dynamics in the absence of grazing is, by the calculations summarised in Table 10.1, about double that in the presence of grazing (CV = 23% against CV = 39%). That conclusion is tentative because of a difference between Systems A and C in the calculation of pasture growth, but it makes sense dynamically. Under this modelled regime of grazing in which the kangaroos are neither preyed upon by people nor share the pasture with domestic stock, the grazing had little effect on the magnitude of fluctuation of pasture biomass (Table 10.1: $s = 121$ against $s = 115$) but reduced the long-term average by about 40%. That percentage is tentative also, for the same reason noted in relation to the coefficients of variation above.

Lags in the system's dynamics

Any lag between cause and effect within the dynamics of a system will reduce the centripetality of the system (May, 1975, pp. 95-98). One of the firmer findings of this study, and one of the more interesting, is the rapidity with which effect follows cause. Pasture responds to rainfall within a few days and its total biomass is determined largely by rainfall over the previous six months (Chapters 3 and 4). A burst of pasture growth enhances the condition of kangaroos within a month (Chapter 9). Rate of increase of kangaroos responds to rainfall within six months, and most of that lag is in detection time (Chapter 8). The kangaroos' dispersion by soil type responds to rainfall within a few weeks (Chapter 7). Red kangaroos conceived within two weeks of the breaking of the 1982-83 drought but western grey kangaroo did not (Chapter 9).

Few lags in this system exceed six months. Most are measured in days or weeks rather than months or years. Thus time lags are likely to have scant influence on the dynamics of this system.

Medium-term dynamics

Figure 10.1 shows that the year-to-year rainfall of this climatic zone takes the form of high-amplitude, high-frequency fluctuations. The pasture, made up mostly of annuals that germinate fast and grow rapidly after rain, dying back almost as rapidly when soil moisture declines, generates a trace of similar high-amplitude, high-frequency fluctuations whether grazed or ungrazed. Grazing by kangaroos lowers the long-term mean of pasture biomass but has little effect on either the relative amplitude or the frequency of its fluctuations.

Although kangaroo populations react dynamically to pasture biomass with little lag the resultant density trace differs markedly from that of the rainfall and pasture. Relative amplitude of the fluctuations is about the same but frequency is much lower. Kangaroos may pass through a decade of high densities followed by a decade of low densities, the total rainfall of those two decades being the same. The reason for the difference between the trace of kangaroo density and of pasture biomass lies in the very much faster rates of growth and dieback of the latter. Pasture biomass can increase by a factor of 100 over a year, kangaroos by 1.5. Pasture can die back to a tenth of its starting biomass over a year. The kangaroos might react to that drop by increasing (if the plunge in pasture biomass levelled out above about 200 kg/ha) or they might decline by 50% if pasture bottomed out at 10 kg/ha. The large difference in realisable rates of change between kangaroos and pasture explains the inability of kangaroo density to track faithfully changes in food supply. Rate of increase of kangaroos does that. Their density does not. The trend of density is conservative. Present density is an integration of past rates of increase, not of present conditions. If density is pushed high by a sequence of annual rates of increase tending above zero, it is likely to remain high when conditions return closer to their long-term mean. When pushed low by a drought the density may remain low for some time despite adequate rainfall.

Present density of kangaroos (as against their rate of increase) may bear little or no relationship to present food supply. There is thus a lot of slack in the dynamics of this grazing system because the grazers and the grazed operate on different scales of time.

Short-term dynamics

Figure 10.2 shows simulated rainfall in three-monthly intervals over ten years and two trends of ungrazed pasture biomass generated by that sequence of rainfall. The model is that of System C (Table 10.1) but with kangaroo density set at zero. The biomass traces differ in starting density, one beginning from 100 kg/ha and the other from 800 kg/ha. Despite that considerable disparity in initial conditions the two trends come together within three years.

Figure 10.3 shows the modelled responses of grazed pasture and of kangaroos to that same sequence of rainfall, with high and low starting biomasses of pasture combined with high and low starting densities of kangaroos.

In graph *a* the high starting biomass of pasture (800 kg/ha) was combined

Fig. 10.2. A draw of quarterly rainfall from the Menindee climate over ten years and the expected fluctuations of ungrazed pasture biomass that it would produce. The two replicates which differ greatly in starting biomass come together within three years.

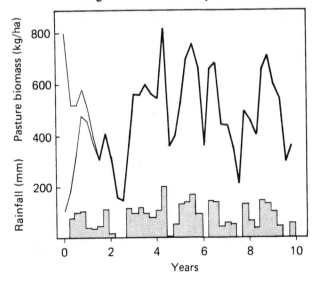

with a high initial density of kangaroos (0.5/ha). Although a drought in the third year reduced the population by 55% it regained its pre-drought density within three and a half years and finished the decade with double its starting density.

Graph *b* commences with a low biomass of pasture (100 kg/ha) and a high density of kangaroos (0.5/ha). The density of kangaroos declined initially in response to the dearth of food, was reduced further by the drought to much the same level as the population of graph *a*, and thereafter the trends of the two were concordant.

Graph *c* started with high pasture biomass (800 kg/ha) and low kangaroo

Fig. 10.3. The run of modelled rainfall presented in Fig. 10.2 is here utilised by pasture grazed by kangaroos. The consequences of four sets of initial conditions are investigated: (a) high initial kangaroo density and high pasture biomass, (b) high kangaroos and low pasture, (c) low kangaroos and high pasture and (d) low kangaroos and low pasture.

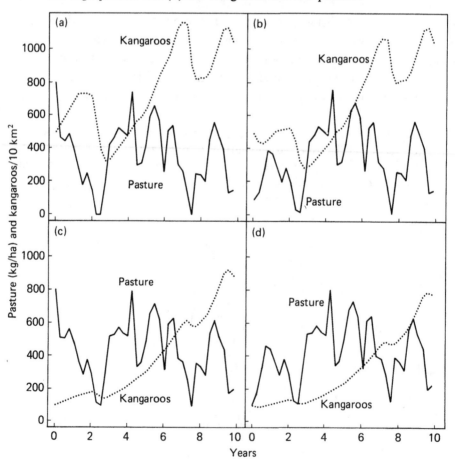

density (0.1/ha). The low grazing pressure allowed the pasture to remain above 100 kg/ha during the drought, and the kangaroos increase almost continuously. By the end of the ten years they had increased by a factor of nine.

Graph *d* combined a low starting density of pasture with a low density of kangaroos. The trends were similar to those of graph *c* after an initial period of readjustment.

The trends of Fig. 10.3, taken together, indicate how the present density of kangaroos is influenced by the previous densities of kangaroos and by previous levels of pasture biomass. The history of pasture biomass has little effect. Whether biomass has been high or low in a given year, the memory of that event, as reflected in the density of kangaroos in subsequent years, will be lost within three or four years. But whether kangaroo density has been high or low in a given year is an important determinant of subsequent density. Its influence may linger for ten years or more.

10.45   *The two main species of kangaroos*

The system's dynamic behaviour was explored in the previous section by modelling the relationships between weather, vegetation and red kangaroos. The system contains more herbivores than that: western grey kangaroos, rabbits and insects also contribute to offtake of pasture within the park. However the rabbits were not a significant influence on pasture during the course of this study (see Chapter 4) and our data on offtake by insects is not sufficient to allow any generalisation.

The western grey kangaroo is the other dominant herbivore on the park. Its diet differs in detail from that of the red kangaroo but there is broad overlap (Chapter 5). Its use of habitat differs from that of red kangaroos but again there is broad overlap. Its reproduction is more seasonal than is the red's and its rate of recruitment is probably lower (Chapter 9). It moves around about as much as does the red kangaroo (Table 7.3). Its rate of food intake as a function of pasture biomass takes the same form as the red kangaroo's but it is less efficient at finding food when pasture is sparse (Chapter 6).

The point is made in Chapter 9 that the western grey kangaroo appears somewhat less adapted to this environment than is the red kangaroo but the similarities in the ecologies of the two species overshadow the differences. This is reflected in the observed dynamics of the two species on the park and the surrounding sheep stations between 1973 and 1984 (Fig. 8.2). Western greys tended to increase when reds increased and declined when reds declined. Their rate of increase was marginally lower during favourable conditions and their decline during the 1982-83 drought was marginally steeper, but dynamically the two species were doing much the same thing. We conclude therefore that their effect on the dynamics of the system is similar, that a system containing red and grey kangaroos would behave in a manner similar to that modelled for red kangaroos alone. We suspect that their coefficients of competition are high and that an artificial reduction in the density of one would be compensated to some extent by an increase in the other.

10.46   *Kangaroos and sheep*

Kinchega National Park and the abutting sheep station of Tandou received, save for minor differences, much the same weather over the period of the study. One would expect therefore that the ups and downs of pasture biomass on the two study sites would be roughly parallel. So they were, but additionally they were almost concordant, the two trend lines having to be drafted carefully lest one obliterated the other on the diagram (Fig. 4.3). That unexpected result can be interpreted in two ways: that weather so dominated pasture biomass that it over-rode the difference (kangaroos on Kinchega, kangaroos and sheep on Tandou) between those two grazing regimes; or that those two grazing regimes, differing in kind as they did, produced a similar offtake of pasture. Luckily the offtake was measured independently (Chapter 4) and it is high enough to indicate that offtake itself was an important influence on the standing crop of pasture (see the pasture budgets of Fig. 4.11). At a given density of herbivores, offtake of pasture will increase as pasture biomass increases until a threshold is reached where any further increase of biomass elicits no further increase in offtake. The animals are then taking in a satiating diet. A threshold biomass of around 300 kg/ha is suggested for red kangaroos and sheep by the functional response curves presented in Chapter 6. It may be higher for western grey kangaroos (Fig. 6.1). Minor differences in offtake between Kinchega (kangaroos only) and Tandou (kangaroos and sheep in varying proportions but 40:60 on average) are explicable largely in terms of differing grazing pressures expressed in kangaroo units (Fig. 4.8). There are anomalies that we do not understand, particularly in 1981 and 1984, but the general effect seems clear enough.

Sheep can be transformed algebraically into kangaroos according to how much they eat. At its most primitive, this translation requires no knowledge of what they eat, only how much. The scaling is made according to body weight raised to the power of 0.75, the technique used in Chapter 6 to compare the intake of sheep, rabbits, red kangaroos and western grey kangaroos against a common scale of metabolic weight. By this criterion a sheep is worth about 1.5 kangaroos (Chapter 4 favours 1.6; Chapter 6 prefers 1.4).

A herbivore mix of around 60% sheep and 40% kangaroos cropped about the same amount of pasture as did another mix of herbivores comprising only kangaroos (Chapter 4). However that influence will not feed back to the dynamics of either kangaroos or sheep until pasture biomass is grazed below or dies back below about 300 kg/ha. Even if they were eating exactly the same mix of plants and plant parts there cannot be, by definition, any immediate competition at higher levels of pasture biomass. Since the diets of these species overlapped considerably (Chapter 5), although that proportional overlap lessened when food becames scarce (Fig. 5.6), they influence each other's food supply for much of the time and compete at biomass levels below about 300 kg/ha.

Ecological effect of sheep on kangaroos

The model used previously to explore the relationship between pasture biomass and red kangaroos is here elaborated to examine the likely effect of introducing sheep into this system. Sheep are introduced as kangaroo equivalents but without the numerical value of the trade-off being specified or even guessed at. We seek only the effect on long-term pasture biomass and kangaroo density of injecting into the system a number of sheep that eats a fixed proportion of the food sought by the kangaroos. The results in Table 10.2 suggest that this has little effect on mean pasture biomass but that the kangaroos will drop in density as part of their food supply is pre-empted by the other species. By these calculations, if sheep are held at a density such that their offtake of pasture components eaten by kangaroos is instantaneously identical to the kangaroos' offtake, the long-term density of kangaroos is roughly halved. This mechanism, together with the harvesting of kangaroos on Tandou, is offered as an explanation of why the density of kangaroos on Tandou is less than half that on Kinchega National Park.

That much does not depart too far from commonsense. What is less obvious is that although sheep have a marked influence on the long-term density of kangaroos, they have a negligible effect on the short-term dynamics of those populations. A fall of rain, the effect of which flows through the pasture to stimulate an increase of kangaroos at a rate of say $r = 0.2$ in the absence of sheep, will stimulate an increase of $r = 0.2$ in the presence of sheep: likewise for a drought triggering a decline. That conclusion is contingent upon the ratio of sheep to kangaroos being held constant as in the simulations summarised by Table 10.2, or the sheep being held at constant density, providing only that either stocking strategy has remained unchanged long enough (about ten years) for mean kangaroo density to have adjusted to it. This effect is signalled in Table 10.2 by the coefficient of variation of kangaroo density among years.

In the simulations summarised there it sits at about 58% regardless of the ratio of sheep to kangaroos. Sheep will have an effect on the dynamics (as against long-term density) of kangaroos only when the density of sheep is manipulated antagonistically to the trend of pasture biomass.

In the real world the effect of sheep on kangaroos is likely to reach a peak of intensity during drought when pasture is at low biomass. At this time the diet of the sheep broadens to include a significant proportion of bluebush whereas kangaroos eat bluebush sparingly and cannot survive on it for long. Sheep are often hand fed with hay or grain at this time in such a way that denies kangaroos a share of it. In the mulga (*Acacia aneura*) shrublands branches of mulga are lopped to feed sheep during droughts, but red kangaroos and eastern grey kangaroos (Griffiths & Barker, 1966), and probably also western grey kangaroos, rarely touch it even when alternative food is almost exhausted. Despite the divergence of diets during drought the absolute competition for food is greatest at this time. The proportions of plant species eaten during drought certainly differ among sheep, red kangaroos and western greys, but of those species shared, albeit disproportionately, there is not enough available to supply the individuals of any species with a maintenance diet. That competition for food is asymmetrical. Sheep seek the pasture species eaten by kangaroos. Kangaroos do not, as far as we can discover, eat any species shunned by sheep. Sheep eat some species avoided by kangaroos and they benefit also from supplementary feeding denied to kangaroos. Consequently, as pasture biomass is reduced during a drought by dieback and

Table 10.2. *Modelled mean density and variability between years of pasture and kangaroos over 100 years. The sequence of weather is the same for all runs. The amount of pasture eaten by sheep is instantaneously in fixed ratio with the amount eaten by kangaroos (see text).*

Conditions	Pasture kg/ha	s	CV%	Kangaroos n/ha	s	CV%
Kangaroos only	286	94	33	0.434	0.258	59
Sheep eat half as much as kangaroos	284	94	33	0.295	0.172	58
Sheep eat the same as kangaroos	283	94	33	0.223	0.129	58
Sheep eat twice as much as kangaroos	·282	94	33	0.150	0.086	57

grazing, the density of kangaroos is reduced by heightened mortality and curtailed breeding. Sheep density declines also but not to the same extent. The 1982-83 drought reduced the number of adult kangaroos by about 50% on Kinchega and on Tandou whereas adult sheep mortality on Tandou was only about 16% over the same period. Thus the competitive effect of sheep on kangaroos is maintained well into the drought when the reciprocal effect is weakened by a steep fall in kangaroo numbers.

Ecological effect of kangaroos on sheep

Data presented in this book suggest that the competitive effect of kangaroos on sheep, working through a shared food supply, is likely to be considerable, but we do not have the experimental results to quantify it precisely. Although competition is likely to be most intense during droughts, the modelling reported in Section 10.42 suggests that it will also have a long-term average effect. Table 10.1 suggests that the mean biomass of pasture is reduced by an average of about 40% if it is grazed by an unharvested population of red kangaroos. Just as Table 10.2 indicates that an injection of sheep into a pasture-kangaroo system will reduce the long-term density of kangaroos, the same effect may be expected of kangaroos introduced into a pasture-sheep

Fig. 10.4. Modelled mean offtake of kangaroos per km² per year in the absence of sheep, graphed against the mean instantaneous harvesting rate multiplied by 100, per year. Each point results from an independently simulated century of Menindee weather. The trend is fitted by eye.

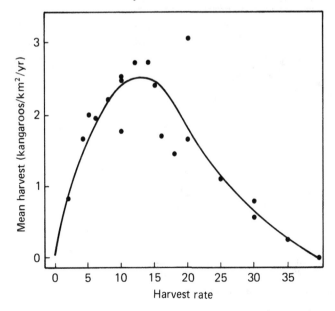

system. The relative effects are likely to be asymmetrical during drought, that of kangaroos on sheep being less than the effect of the sheep on kangaroos, but both effects will be far from trivial.

10.47   *Effect of harvesting on kangaroos*

The System C model of Table 10.1 was used to explore the likely effect of harvesting on red kangaroos. A tracking strategy of harvesting was simulated, the number shot per year rising as density rose and falling as density fell. That was imposed on the model by holding the harvesting rate constant, the same proportion of the population being shot each year whether density was high or low.

Figure 10.4 shows the average effect of a number of 100-year simulations. Each point is the result of one such run graphed as mean annual harvest against the instantaneous harvesting rate. A trend line sketched through these points indicates that yield is maximised over a long period by harvesting between 10% and 15% of the population each year. That value is likely to be somewhat lower for western grey kangaroos. Of particular interest is the scatter around the trend. Each point is generated by a sequence of weather with a common mean and variance but, although each run represents 100 years, that is too short a time to generate a stable average result for a given harvesting rate.

Fig. 10.5. The simulated offtakes of Fig. 10.4 are here graphed against the average 100-year densities of kangaroos enforced by those differing rates of harvest. The trend is fitted by eye.

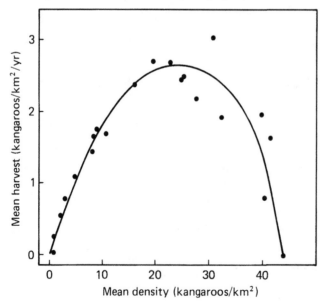

Figure 10.5 differs from the previous figure in that harvest is now plotted against the mean density over 100 years that a given harvesting rate enforces. The highest long-term offtake, that corresponding to a harvesting rate between 10% and 15% per year, depresses the kangaroo population to between 60% and 70% of its unharvested mean density. Again the scatter around the trend is high, indicating that the trend is a guide, not a prescription.

The presence of sheep will depress the density of kangaroos, and hence offtake for a given harvesting effort, but they should not affect the 10%-15% identified as the interval of harvesting rates generating a maximum long-term yield.

10.5   What regulates kangaroo populations?

In this section the information provided previously is re-examined in the context of current ecological theory on the regulation of animal populations. We deal first with the system comprising rain, pasture and kangaroos, the dynamics of those kangaroos being influenced neither by harvesting nor by a sharing of their food with sheep. Kinchega National Park provides our example.

The naive response to the question posed by the section heading is that the kangaroos are not regulated because they can (and do) go up and down by large amounts with wearying frequency. The reasons for that instability are not difficult to find: rate of increase and decrease of kangaroos is determined by pasture biomass, and pasture biomass is largely determined, with remarkably little lag, by rainfall. Rainfall is erratic, the swings of pasture biomass are likewise erratic, and the swings in kangaroo density are generated and maintained by the swings in pasture biomass.

But that is not the whole story. The kangaroos themselves have an effect on pasture that can easily be missed at first glance. The changes in pasture biomass are so rapid and of such magnitude that one might assume that no vertebrate herbivore could have any real effect on them. Often this is so. When kangaroos are at low density they cannot do much to a burst of germination and growth. Whether they are at high or low density they are unlikely to have much effect on the initial rapid die-back of pasture when a drought bites. But appearances are deceptive. Despite often being ambushed when at high density by a drought that cuts the food from under them, and being trapped inappropriately at low density when a burst of pasture growth provides enough food to maintain a hundred times their number, they are not merely spectators of this game. They tend to clip the tops off the peaks of pasture production and deepen the troughs. Over a long run they depress pasture biomass to about 60% of its ungrazed mean level. If they did not do that the kangaroos would tend, although fluctuating in density, to increase continually. The kangaroos are not totally at the mercy of this fluctuating environment. The feedback loop comprising more pasture, more kangaroos, less pasture, fewer kangaroos, imposes a centripetality upon the system. Pasture biomass in the absence of vertebrate herbivores is also centripetal because pasture has its own feedback which depresses growth as biomass increases. The kangaroo-pasture loop is often weak because the ratio of kangaroos to pasture is seldom where it should be according to the rigid rules governing a system with constant external input, but it is strong enough in the long run to change apparent chaos into a weakly disciplined interaction.

This system differs in kind from one entirely at the mercy of the weather and it differs in degree from one operating in a stable environment. It can be summarised in terms of three critical values: 200, 300 and 500 kg/ha of pasture:

(1) In the absence of kangaroos and other vertebrate herbivores the pasture would hold to a long-term average of about 500 kg/ha.

(2) Kangaroos increase when pasture is above about 200 kg/ha and decrease when it is below that level.

(3) In the presence of kangaroos the pasture will assume a long-term average of about 300 kg/ha, the mean having been reduced to that level from 500 kg/ha by grazing.

The last two conclusions appear to contradict each other: since the pasture settles to a mean level at which kangaroos increase, the system cannot be centripetal. That objection would be valid if rainfall were constant from year to year. In this environment it fluctuates, and that fluctuation acts upon an asymmetry in the kangaroos' response to pasture biomass. The trend of their rate of increase on pasture biomass is convex upward and asymptotic. Hence a given decrement to pasture biomass depresses the kangaroos' rate of increase more than the same increment enhances it. Table 10.3 illustrates that by showing what would happen to mean rate of increase over two years when pasture biomass, although averaging 300 kg/ha over the period, differed between the two years. The greater the variation in pasture between years the lower will be the rate of increase of the kangaroos over those two years.

If the environment were stable and the rainfall were constant from year to year this system would settle to a relatively constant density of kangaroos utilising a relatively constant biomass of pasture, the level of pasture biomass being determined more by grazing pressure than by the growth characteristics of the plants. That equilibrium point represents an accommodation between the grazers and the grazed, a purely mechanical consequence of the herbivores increasing when pasture biomass is above a certain level and decreasing when it is below that level. Pasture biomass would settle on that critical level and the kangaroos would settle on a density sufficient to keep pasture grazed down to precisely that level. If either were displaced from its equilibrium level the other would first move in the opposite direction and then the two would converge upon their original values. Those density and biomass values map the position of the theoretical equilibrium between pasture and kangaroos in a constant environment.

In a fluctuating environment the rules change. The point around which the plant-herbivore system is moved by a fluctuating rainfall is not the equilibrium of the constant environment. It is displaced from that position in the direction of more pasture and fewer kangaroos. The greater the environmental variability in time the farther is this centre of centripetality displaced from the 'equilibrium point'. The calculations summarised in Table 10.1 (System C compared with System G) suggest that the 40% coefficient of variation in the annual rainfall of this system shifts the centre of centripetality to where pasture is on average 30% higher and kangaroo density is on average 40% lower than those of the 'equilibrium point'. The centre of centripetality is real and measurable. Its coordinates are the long-term mean density of kangaroos and the long-term mean biomass of pasture.

10.51   Modes of regulation

A diversion is required at this stage to note some of the background to current controversies about regulation of populations. Because some of the

Table 10.3. *Rate of increase (r) of red kangaroos to be expected at given levels of pasture biomass (V) in kg/ha. Each line gives V in two consecutive years, their mean, the two rates of increase corresponding to the two levels of pasture, and the mean of those two rates. As the pasture biomass varies increasingly around the common mean of 300 kg/ha the mean rate of increase of kangaroos declines.*

V_1	V_2	Mean V	r_1	r_2	Mean r
300	300	300	0.155	0.155	0.155
250	350	300	0.052	0.227	0.140
200	400	300	−0.093	0.278	0.093
150	450	300	−0.300	0.314	0.007
100	500	300	−0.593	0.340	−0.127
50	550	300	−1.009	0.357	−0.326

$r = -1.6 + 2(1 - e^{-0.007V})$.

controversy has more to do with the meanings of words than with the interpretation of events it will be necessary to explain some of the subject's jargon.

Intrinsic and extrinsic regulation

Regulation of populations can be divided into two categories according to the mechanism of regulation. The first category is 'intrinsic regulation' where a population's rate of increase is lowered by spacing behaviour of some sort as density rises. In a constant environment that mechanism enforces an equilibrium. Intrinsic regulation (or self-regulation) is a property of the animals — its program is carried on their genes — but the level of the equilibrium may reflect interaction between the program and the environment. For example the equilibrium level of animal density may be high where food is abundant and low where food is scarce. The dynamics of an intrinsically regulated population can be modelled adequately in terms of its rate of increase and its density to produce what is called a single-species model.

In contrast, the equilibrium of 'extrinsic regulation' is enforced by a relationship between the animals and a limiting resource. The commonest example, but not the simplest, is a population of animals feeding on one or more populations of plants. The density of the animals influences the rate of increase of the plants, and the density of the plants influences the rate of increase of the animals. That interaction also leads to an equilibrium in a constant environment but it is a mechanical equilibrium that would occur irrespective of genetic programming.

The two modes of regulation can be separated conceptually by asking the question: does the density of a population *instantaneously* determine its rate of increase? If so the animals are doing something to each other and regulation is therefore intrinsic. If rate of increase at a given instant is determined only by the amount of a resource in front of each animal's nose and has nothing to do with the density of animals at that instant, the animals have no instantaneous effect on each other and regulation is therefore extrinsic.

The two modes of regulation need not exclude each other. Either can alone regulate a population but there is nothing to ban a bit of both. Thus the identification of an instantaneous effect of density on rate of increase indicates that intrinsic regulation is operating, but extrinsic regulation may be occurring in tandem with it. The discovery that density has no instantaneous effect on rate of increase identifies any observed regulation as purely extrinsic.

Density-dependence and density-independence

These terms are introduced here mainly because they tend to get mixed up with those of the previous dichotomy. A population's dynamics may be described as 'controlled by density-dependent factors' or simply as being 'density-dependent' when rate of increase is predictable from density. The dynamics may be described instead as 'density-independent' when there is no necessary correlation between rate of increase and density. The first term is often used as a synonym for intrinsic regulation, the second for extrinsic regulation, and that causes much confusion. Both intrinsically and extrinsically regulated populations can exhibit density-dependence. The intrinsically regulated population exhibits it instantaneously because a change in density is the causal trigger to a change in rate of increase. The extrinsically regulated population is density-dependent only when viewed over an interval of time. An example would be a population of deer for which the limiting resource is grass. At any given moment the population's rate of increase is determined only by the biomass of grass, but that biomass is influenced by the previous history of grazing. Hence present rate of increase may well be predicted from the integral of density from the present back to some time in the past, and in that sense the dynamics of the population are said to be density-dependent. No harm is done thereby unless the demonstration of this form of density-dependence is used to imply that the population is self-regulated. That can lead to misconceptions about the evolution of population processes, and it can also lead to blunders of management because the ecological relationships and mechanisms have been misidentified.

Regulation of kangaroos
In this environment the driving variable is rainfall. Kangaroo populations go up and down according to the effect of rainfall on plant growth. Plant biomass is determined by the rainfall over the previous six months and by grazing pressure over the same interval. An individual kangaroo's intake reacts instantaneously to a change in pasture biomass, its probability of survival reacts to pasture biomass over a few months, and its probability of reproducing reacts to it almost instantaneously for red kangaroos and seasonally for grey kangaroos. In the sense that past grazing pressure influences present plant biomass, the dynamics of the kangaroos may be described as density-dependent, if one so desires. The danger in doing so is that it might create the false impression that there is a causal relationship between present rate of increase and present density. Similarly, because there is no causal connection between present rate of increase and present density, the dynamics of the population could fairly be described as density-independent. That would not be a good idea either because it could be taken to indicate that the system lacked

centripetality. Both terms create more problems than they solve and will not be used further.

The evidence presented in this book suggests that rate of increase of kangaroos in this environment is predicted best by pasture biomass which in turn is predicted best by rainfall. These predictions reflect direct and causal relationships. There is additionally a weak effect of grazing on pasture biomass that turns out, paradoxically, to be very important. It imposes a centripetality upon the kangaroo-pasture system, the centre of centripetality being displaced from the equilibrium point of a constant environment. The kangaroos are regulated extrinsically. We found no evidence of an additional intrinsic regulation but it could be argued fairly that we did not look for it very hard.

This section began with the question 'what regulates kangaroo populations?' because that is a question often asked. In fact it is not a particularly useful question because it reflects an implicit acceptance of the single-species paradigm in which a population is seen as regulating itself. The question limits the breadth of the answer. The more useful question is 'by what means does a kangaroo-pasture system maintain centripetality?' That allows a full exploration of interactions within the system as a whole, rather than from the viewpoint of but one of its components, and that has been the approach used in all sections of the chapter other than this one.

Hairston, Smith & Slobodkin (1960; see also Slobodkin, Smith & Hairston, 1967) hypothesised that, unlike producers, predators and decomposers, 'herbivores are seldom food-limited, appear most often predator limited, and therefore are not likely to compete for common resources.' Whether that is true in general, and two recent analyses of a large number of field experiments reached opposite conclusions, one agreeing, the other disagreeing (Connell, 1983; Schoener, 1983, 1985), it is clearly not true for this system.

May (1973) advanced the then novel idea that complex systems are likely to be no more stable than simple systems and on average will be less so. This system conforms with his conclusion in that the vegetation biomass in the absence of kangaroos has a lower coefficient of variation between years than it does in their presence.

10.6   Ecological questions remaining to be answered

The questions posed by this study were selected according to prior guesses as to which would be important to, and which would be trivial to, an understanding of the system. The study provided some answers, generated further questions, and ignored a number of questions that would have been asked had we the time and resources to answer them. In this section a number of ecological questions are flagged that might profitably be answered by a subsequent study.

(1) Our findings on the ecological effect of kangaroos on sheep and of sheep on kangaroos are suggestive rather than conclusive. The logical next step is a two-factor experiment in which sheep and kangaroos are combined at different densities, including zero. Pasture biomass and growth, pasture composition, herbivore condition, growth and reproduction, could then be compared between treatments.

(2) This study used the simplifying assumption that there is no long-term trend in species composition of the vegetation. That needs to be checked by monitoring *grazed* vegetation. The trend in exclosure plots does not answer the question.

(3) We were unable to calculate growth, die-back, offtake and biomass of pasture over the 640 km^2 studied within intervals shorter than three months. This provides a distorted picture of continuous processes. A study of pasture dynamics over much shorter time intervals would sharpen the conclusions presented here.

(4) Many of our data come as estimates of total pasture biomass rather than pasture composition and quality. That reflects an initial assumption that changes in the density of the herbivores would bear a much closer relationship to changes in the amount of pasture than to changes in the quality of the pasture. We guessed that calories would be more important than nutrients, and in retrospect do not regret that punt, but future studies on the effect of food quality on herbivore dynamics will refine our conclusions.

(5) We have inconsistent data on the recovery of the kangaroo populations after the 1982-83 drought. The aerial survey figures (Chapter 8) suggested that red kangaroos continued to decline for some time after the drought broke. The data on condition and recruitment (Chapter 9) suggested, in contrast, that red kangaroos reacted rapidly to the improvement in nutrition. By the latter findings the populations should have first held level and then increased. Apparently either the aerial survey estimates were faulty over that period or there was further mortality of adult kangaroos anomalous with respect to their nutrition, for which we have no evidence. Further study is needed.

(6) We have concluded from rather sketchy data that spacing behaviour is an unlikely determinant of kangaroo dynamics and hence that intrinsic regulation is not a feature of their ecology. We would be more comfortable with that conclusion were it checked by an independent study seeking specifically to falsify it.

(7) The offtake of pasture by insects may be a sporadic but important influence upon pasture biomass, and hence upon the ecology of this system.

The possibility should be checked.

(8) We touched upon the effect of harvesting on kangaroo populations but lack precise data on the relationships among rate of increase, harvesting rate and pasture biomass. A future experimental study would fill that gap and add to our knowledge of the effect of manipulation upon the system.

(9) More information is needed on the ecological separation and overlap among sheep, red kangaroos and grey kangaroos, particularly in terms of habitat selection and diet.

10.7   Conclusions

The strategy of this study was to determine the existence and then the form of the interrelationships within this system, to describe them numerically, and finally to explore the ecological consequences of those relationships.

10.71   *Elements of the system*

The pasture layer comprises mainly annual and ephemeral species and some short-lived perennials. The long-lived perennials are waist-high shrubs loaded with salt and eaten only when there is little else to eat.

The growth of pasture is closely predicted by rainfall, much less so by either soil type or month of the year. Since seed germinates rapidly in response to rainfall and the plants grow rapidly thereafter, total pasture biomass is predictable at any time from rainfall over the previous few months. Although rainfall is the driving variable it is not correlated with anything, being highly variable from month to month and from year to year. Hence pasture biomass is highly labile and may vary by up to a hundred times over a couple of years.

The amount of pasture a kangaroo consumes in a day is determined almost exclusively by its body weight and by the pasture biomass on that day. Its rate of consumption is approximately constant when pasture exceeds about 300 kg/ha but declines steeply when pasture falls below that level. The relationship differs in detail between red kangaroos and western grey kangaroos, the former being more capable of finding food when food is scarce. Kangaroo populations increase when pasture biomass is above about 200 kg/ha and decrease when it is below that level.

The system contains two negative feedback loops. Pasture grows less vigorously at high biomass and stops growing at very high biomass. We call this the pasture biomass loop. The second links the pasture to the kangaroos: the more pasture, the more each kangaroo eats and the more kangaroos are recruited into the population; which provides higher grazing pressure, less pasture and then fewer kangaroos. That is the plant-herbivore loop.

10.72 *Behaviour of the system*

The ecological consequences of those relationships were explored by linking their measured coefficients with those of the weather and then running them over simulated time. That exercise indicated that although the short-term changes in pasture biomass and kangaroo density appear chaotic they are tightly deterministic. The apparent chaos is generated by the fluctuating weather.

There is a higher rationality in the long term. The dynamics of the pasture-kangaroo system, although being tossed around by the weather so much that the system never stands still, are nonetheless dampened by feed-back loops that impose centripetality. The simulations indicate that the dynamics of ungrazed pasture would be centrifugal without the pasture biomass loop. Without that loop but with grazing by kangaroos the system is centripetal. With both the pasture biomass loop and grazing by kangaroos the system is even more centripetal.

If sheep were introduced into such a system, their density being controlled by people rather than by environment, the density of kangaroos would drop by about the same proportion as that by which their food supply is pre-empted by the sheep. The level of pasture biomass would be little affected in the long term, the dynamic adjustment being made by the herbivore population rather than by the pasture. Culling of kangaroos acts differently. The density of kangaroos is reduced and average pasture biomass increases.

The centre of centripetality, that value of pasture biomass and kangaroo density around which the state of the system is moved non-directionally by erratic rainfall, is not the equilibrium point upon which the system would converge were the rainfall constant from year to year. It is displaced from that point in the direction of more pasture and fewer kangaroos. The magnitude of the displacement is proportional to the coefficient of variation of annual rainfall. The reason for the displacement is that rate of increase of kangaroos is affected more by a decrease in pasture than by an increase. The greater the annual variation in rainfall (and hence in pasture) around a given mean, the lower is the kangaroos' rate of increase and hence the fewer kangaroos and the more pasture. Kangaroos will increase whenever pasture exceeds about 200 kg/ha but their rate of increase averaged over a long period will be zero when pasture biomass averages 300 kg/ha over the same period. That paradox is resolved by the fluctuating environment.

11

Options for management of kangaroos

NEIL SHEPHERD AND
GRAEME CAUGHLEY

11.1   Introduction

There are only three ways that we can manage a population of wildlife. We can control it or even exterminate it, we can utilise it to provide a continuing yield, or we can cherish it for its intrinsic worth. The decision as to the right option will be determined mainly by what we consider the appropriate use of the land on which the animals live. A population of kangaroos living in a paddock of wheat will not be viewed in the same way as one living in an unutilised desert.

In this chapter we explore the options available for managing kangaroos on two classes of land — national parks and nature reserves, the perceived primary function of which is to conserve, and rangelands whose perceived primary function is to produce. Most people agree on the primary function of those classes of land, but when those broad functions are specified more fully they reveal an underlying layer of secondary objectives which may conflict with each other. These in turn, depending on the ranking assigned among them, generate a range of procedures whereby kangaroos might best be managed.

11.2   Management within national parks

The appropriate management is determined less by what is happening on the ground than by perceptions of what the national park is for. We will trace the history of these changing perceptions to show how they determine the 'appropriate management' of the soil, plants and animals within the parks. After that wide excursion we will come back to kangaroos.

11.21   *The national park idea*

There is a sameness the world over to the ideas of what national parks are for but that is a comparatively recent result of convergent evolution and mimicry. The national park idea has two quite separate philosophical springs whose streams did not converge until about 1950. The first is American, exemplified by the U.S. Act of 1872 proclaiming Yellowstone as the world's first national park. The land was to be

> 'reserved and withdrawn from settlement, occupancy, or sale under the laws of the United States, and dedicated and set apart as a public park or pleasureing [*sic*] ground for the benefit and enjoyment of the people. . .'

The intent was to preserve great scenery, not animals or plants. Public hunting and fishing were at first entirely acceptable (Houston, 1982; Schullery, 1984).

The second spring is 'British colonial'. Graham (1973, p.55) described its essential features as the Crown asserting right of ownership over game animals, the setting aside of large tracts of land as game reserves, and the prohibiting of hunting thereon without a permit. Permits were available only to privileged persons. The great national parks of Africa grew out of these game reserves, some physically and the others philosophically. Wildlife was the primary concern. Scenery came second if at all. The timing of the shift from game reserve to national park varied across Africa. The first was Kruger National Park established in 1926 on a game reserve proclaimed in 1898. Kenya's first was established in 1946 on the Nairobi common. Many African parks evolved from game reserves *in situ*. The point at which they stopped being one and started being the other is often difficult to define precisely.

The first national park in Australia, and the second in the world, was dedicated in the then Colony of New South Wales in 1879. Essentially it followed the American model, but there is some evidence that the founders had in mind also the commons-land 'parks' on the outskirts of mid-nineteenth century London (Pettigrew & Lyons, 1979). Australian national parks founded subsequently followed the American model closely. Most are recent, few dating back beyond 1950.

All national parks established for forty years or more have had their goals modified, often several times, and the way they were managed changed accordingly. The national park idea has not remained static.

Alternative Goals
National park philosophy is now international and ideas spread rapidly. We list the more influential fashions in park theory, roughly in order of appearance. They are not mutually exclusive. Although one of them can usually be identified as the dominant objective for a given national park at a specified time, new objectives accrete upon previous objectives as often as they replace them.

(1) *The most important objective is the conservation of scenery and of nice animals.* The aim translated into restricting roads and railways and attempting to exterminate the carnivores.

(2) *The most important objective is the conservation of soil and plants.* This aim was a direct consequence of the rise of the discipline of range management in the U.S.A. during the thirties. Its axiom was (and still is) that there is a 'proper' plant composition and density. Enough herbivores were to be shot each year to hold the pressure of grazing and browsing at the 'correct' level. An ecosystem could not manage itself. If left to its own devices it would do the wrong thing.

(3) *The most important objective is the conservation of the physical and biological state of the park at some arbitrary date.* In the U.S.A., South Africa and Australia that date marked the arrival of the first European to stand on the land.

(4) *The most important objective is the conservation of representative examples of plant and animal associations.* The wording has been cribbed from Bell's (1981) definition of the function of national parks in Malawi, but the objective underlies the management of many national parks in many other countries.

(5) *The most important objective is the conservation of 'biological diversity'.* This catch phrase had two meanings. It was sometimes used in the sense of 'species diversity', a contribution to ecological methodology by MacArthur (1957, 1960) whereby the information-theory statistic of Shannon and Wiener could be used to estimate the probability that the next animal you saw would differ at the species level from the last. The statistic is maximised when all species have the same density. Within park management the idea translated as 'the more species the better'. The second meaning dealt with associations rather than species, the more diverse a set of plant associations the better the national park. Porter (1977), for example, defined the objectives of the Hluhluwe Game Reserve in Natal as 'To maintain, modify and/or improve (where necessary) the habitat diversity presently found in the area and thus ensure the perpetuation and natural existence of all species of fauna and flora indigenous to the proclaimed area.'

(6) *The most important objective is the conservation of 'genetic variability'.* The phrase can be defined tightly and usefully (e.g. Frankel & Soulé, 1981), but within the theory and practice of park management it lacked focus. It was tossed around with little or no attempt to define or understand what it meant, whether the variability sought was in heterozygosity, in allelic frequency, or in phenotypic polymorphism. In practice it again translated into 'the more species the better'. But it was seldom used in practice; it was more

prevalent at international conferences of park theoreticians than in the parks. Its underpinning was the theory of island biogeography, another set of powerful ideas owing much to MacArthur (MacArthur & Wilson 1963, 1967).

(7) *The most important objective is the conservation of biological processes.* This idea is recent and, as we show in the next section, differs in kind from the six previous objectives. Frankel & Soulé (1981, p. 98) express it thus — 'the purpose of a nature reserve [in which category they include national parks] is to maintain, hopefully in perpetuity, a highly complex set of ecological, genetic, behavioural, evolutionary and physical processes and the coevolved, compatible populations which participate in these processes.' Don Despain (quoted by Schullery, 1984, p. 71) puts it more succinctly — 'The resource is *wildness*'.

Processes or states?

All of the objectives listed above pertain to conserving something. The first six identify biological states that are to be conserved. The seventh identifies not the biological states but the biological processes as the appropriate target of conservation. At first glance Frankel's and Soulé's 'purpose of a nature reserve' appears also to require the maintenance of states because it refers to the conservation of populations. But populations are not states in the sense that plant associations are states. A plant association has a species composition. Its component populations must have a ratio of densities one to the other that remains within defined limits. If those limits are breached the plant association has changed into another plant association. A population has no such defined structure. Age distribution, sex ratio and density or size need not enter the definition of population.

The management of a national park will be determined by whether the aim is to conserve biological and physical states by suppressing processes or whether it is to preserve processes without worrying too much about the resultant states. Specifically, there are three options :

(1) If the aim is to conserve specified animal and plant associations that may be modified or eliminated by wildfire, grazing or predation, then intervene to reduce the intensity of wildfire, grazing or predation.

(2) If the aim is to give full rein to the processes of the system and to accept the resultant, often transient, states that those processes produce, then do not intervene.

(3) A bit of both — if the aim is to allow the processes of the system to proceed unhindered unless they produce 'unacceptable' states, then intervene only when unacceptable outcomes appear likely.

11.22   *The culling controversy*

Whether or not one manipulates the density of animals within a national park depends on the goals and aspirations for the park. Those goals are value judgements not open to logical disputation. However once the goals have been agreed upon the management activities selected to achieve them may be appropriate or inappropriate, that judgement being technical and open to reasoned debate (Bell, 1983).

One such activity is the continuous or sporadic culling of animals to stabilise their density at a level consistent with the goals of the park. Positions tend to be polarised on this issue and various versions of them can be found in two sets of papers collated by Jewell, Holt & Hart (1981) and by Owen-Smith (1983a). They can be summarised as:

> Herbivores will overpopulate a national park if left to themselves and this will result in the system breaking down to a lower equilibrium. Culling within a national park is necessary and beneficial.
> If herbivores are not culled within a national park the system will remain intact; if they are culled the system's ecological defense mechanisms will be weakened. Culling within a national park is unnecessary and harmful.

The explanation of the the lack of consensus is to be found in the nature of the two positions. They are not testable hypotheses because the opposing sides cannot agree on a crucial test. Neither are they scientific theories in the sense of general explanations of rigorously verified observations. Rather they are scientific paradigms in the sense introduced by Kuhn (1970). A paradigm can be recognised by three characteristics: it is robust in the face of falsified hypotheses and new information, simply expanding to accommodate them without changing its basic structure; the proponents of competing paradigms may talk at each other but seldom listen; and, old paradigms never die, they simply fade away.

11.23   *Carrying capacity*

In temperate and tropical grazing systems, but not in arid-zone systems, violent fluctuation is the exception rather than the rule. Marked perturbations occur occasionally but for most of the time the system's fluctuations are relatively modest and the densities of the animal populations relatively steady. The stability reflects annual rainfall holding to a coefficient of variation below 20% and a strong feedback between the biomass of food plants and the number of animals eating them. We often describe this state by saying that the animals are at or near their carrying-capacity density.

That phrase is unfortunate because it has two meanings. Range managers and pastoralists use it to mean the density of stock that provides a maximum sustained yield. The other kind of carrying capacity is the average density, measured over a long period, of a population that is not culled or harvested. The first version may be termed 'economic carrying capacity' and the second 'ecological carrying capacity'. Attainment of the second is the stated aim of some national parks.

The range managers' idea of what the range should look like is usually imposed by the contrived economic carrying capacity of animal production and influenced little by the ecological carrying capacity of a wild population and a wild vegetation negotiating a mutual accommodation unaided. Consequently they often believe that any variation from economic carrying capacity signals a system out of control, the problem being identified by a plant composition and density differing from the 'optimum'. Range managers called upon to assess vegetation in national parks tend to use as their yard-stick what the vegetation should look like if the park were not a park but a well-run sheep or cattle property in the same climatic zone. The 'carrying capacity' prescribed will be economic. Usually it will be specified in domestic stock equivalents. The prescribed carrying capacity of Umfolozi Game Reserve in South Africa for example is 9.3 AU/km² (Brooks & Macdonald, 1983) where an AU (animal unit) is a cow weighing 454 kg.

Our argument is not against holding densities of herbivores within national parks at economic stock densities. It is against choosing that goal, seldom attainable without continual culling, under the misapprehension that the target density is *the* carrying capacity, that it has an unquestionable ecological and evolutionary legitimacy. A more detailed discussion of the two kinds of carrying capacity, and the confusion that results from their being mixed up, is provided by Caughley (1976b, 1979), Mentis & Duke (1976), Mentis (1977), Eltringham (1979), McCullough (1979) and Houston (1982).

But within the arid zone these problems of theory, fact and interpretation become almost meaningless. The study area has a mean annual rainfall of 236 mm with a standard deviation among years of 107 mm (CV = 45%). Over the period of the study, 1981-1984, the mean annual rainfall was, by chance, also 236 mm, with a standard deviation of 139 mm (CV = 59%) not greatly in excess of its hundred-year value. Thus the weather during the study was essentially what might be expected as a random draw of four years from the climate. Over the same period the mean biomass of the pasture was 324 kg/ha on Kinchega but with a standard deviation among years equal to that mean (CV = 100%), and kangaroo density indices varied among years with a coefficient of variation of 49%. What happened to the system over that period was determined by the variability across years rather than by the mean conditions during those years.

Ecological carrying capacity is, by definition, the long-term mean observed density of animals left to themselves. It is a useful notion for environments that do not fluctuate wildly, where most of the ecology relates to the means of those variables selected to describe the environment. Variances of those variables are less interesting and can often conveniently be ignored as 'noise'. The environment we have described is different. Ecological carrying capacity is still a measurable statistic although a long run of years is required to measure it precisely. The important question is whether, in measuring it, we have retrieved anything illuminating. Perhaps we have, but it is clearly telling us less about the ecology of this system than is the variation around that mean.

Means are important. If mean annual rainfall at Menindee were double what it is, there would be no red kangaroos and very few western grey kangaroos on Kinchega National Park. Eastern grey kangaroos, now rare, would be abundant. Ecological carrying capacity is a function of both the mean and the variance of annual rainfall. Without doubt the environmental mean has a considerable effect on the structure of the system but the dynamics of the system reflect the environmental variance more than the environmental average. The variability can no longer be ignored as 'noise'. It is now the 'signal'.

We suggest that highly variable systems differ both in degree and in kind from slightly variable systems. The threshold above which variance reveals more about ecology than does mean is for us marked by a coefficient of variation of annual rainfall in the region of 30%. Above that, ecological carrying capacity is largely an abstraction, not entirely meaningless but not very useful. We do not argue that the notion is invalid for an arid-zone grazing system. Our objection to it is stronger: that it cramps our understanding of that system.

11.24   *Management of kangaroos within national parks*

In this section we identify which of the various objectives that may be set for a national park would logically call for activities aimed at increasing or decreasing the density of kangaroos within that park. The instruments of manipulation are the wildlife managers' tools of trade: reduction or encouragment of competitors, reduction or encouragment of predators, habitat manipulation including the nurturing or control of fire, the provision of supplementary food, creation or deletion of watering points, and culling.

Culling

We first address the specific issue of whether the ecological system within a park is likely to collapse, and whether there will be irreversible changes, if herbivores are not killed periodically. Pienaar (1983) expresses the fears that he shares with many of his fellow park managers:

> 'It is my firm belief that a laissez-faire or non-intervention attitude [by which he means failure to cull] in the administration of conservation areas in Africa is impractical, fraught with danger and can have unpredictable and even shocking consequences. It may also lead to the extinction of vulnerable or less adaptable species in a destabilized environment, and to the destruction of many unique qualities responsible for the original proclamation of the conservation area.'

Although we know of only one example of kangaroos being culled in a national park (1984, in Hattah-Kulkyne National Park, Victoria) we see culling as an emerging issue in Australia, likely to arise whenever park management is intensified. The subject is hence of more than academic interest.

Because it has happened before there is no doubt that the Kinchega system can be broken down irreversibly to a less complex state. It broke down first about 30,000 years ago when a high proportion of the marsupial fauna was eliminated. The extinctions were biased towards the larger species. We see no reason to search beyond hunting as an explanation of that breakdown while acknowledging that many people seek an explanation in changing climate. Horton (1984) provides a well-reasoned defense of that alternative hypothesis. The system broke down again in the second half of the last century when a further suite of marsupials was eliminated. These were all relatively small species dependent upon ground cover. Reduction of that cover by sheep grazing is the most plausible explanation of those extinctions although the mopping up may have been completed by introduced predators.

Since the potential for breakdown has been amply demonstrated, the possibility of further breakdown is by no means theoretical. However, it must be noted that neither previous breakdown was spontaneous in the sense used in theoretical ecology. The breakdowns resulted not from a temporary environmental perturbation but from the addition of a new and permanent component to the system. Hence they do not suggest that the system has 'multiple equilibrium points'. We ask here whether the system is likely to be broken down by a transient environmental perturbation. That hypothesis was tested for us by the fortuitous intervention of the 1982-83 drought which struck when kangaroos were at high density consequent on a run of favourable years in the seventies. At the time it seemed entirely plausible that the combination of high grazing pressure and low pasture biomass would result in a shift in the kangaroos' diets to the perennial bluebush stratum, which might then have been modified severely or even eliminated. That did not happen. There was certainly a shift of diet towards shrubs (Chapter 5) but the kangaroos could not survive on that food. About 50% of them died whereas the bluebush community remained unmodified. Nor was any species of the pasture layer eliminated so far as we could discover. The evidence suggests that this effect, the vegetation outlasting the herbivores during the drought, is independent of the length and depth of the drought. Kangaroos will starve to death before irreversibly modifying the vegetation of this system. The explanation lies in the shrub stratum being invulnerable to kangaroos (but not to sheep) because they cannot survive on such a diet for long, and in the plants of the pasture stratum being able to survive a bout of unfavourable conditions by withdrawing to the seed bank.

Management of watering points

Whether or not additional watering points should be created within national parks is a hotly debated issue in parts of Africa. Animals tend to concentrate in the vicinity of watering points and modify the vegetation around them. It is not an issue on Kinchega because the park is well watered. Additional watering points would have little or no effect on the ecology of the park.

Effect of fencing

Hanks *et al.* (1981), Laws (1981), Owen-Smith (1983b) and Pienaar (1983) have argued that a fence around a national park will disrupt the dynamics of the animal populations within it by restricting dispersal. Such an effect has been shown for small mammalian herbivores (Krebs, 1971) but not for large herbivores. Caughley & Krebs (1983) maintained on theoretical grounds that such an effect was unlikely, but that is hardly a demonstration. Nor is Owen-Smith's (1983b) computer modelling exercise.

We had intended to test this hypothesis on Kinchega, using the kangaroo populations outside the fence as controls. Unfortunately the experiment was flawed by a leaky fence (Chapter 7). The lack of a marked difference between the dynamics of the populations inside and outside the fence therefore no longer tests that hypothesis.

Management to achieve park objectives

The evidence provided by this study suggests that the ecological system within the park is likely to remain intact in the absence of culling. However, management activities aimed at the kangaroos, including culling, may none-theless be needed to achieve specific objectives laid down for the park, other than that of ensuring the persistence of the system as a whole. We therefore list various objectives that have been advocated from time to time for various parks in various countries, and identify which management activities would be needed to achieve them in this park. Objectives are judgements of value and so we do not burden the reader with our own opinion on which of these are far-sighted, myopic or blind. Our personal value judgements are relevant only if they distort, consciously or unconsciously, our technical judgements. The suspicious reader can check the prejudices of one of us (Caughley, 1983). Those of the other differ, but more in degree than in kind.

These objectives have been suggested for various national parks:

(1) *Stabilise the system*

This objective is not feasible because wide fluctuations in weather are intrinsic to the system. Pasture biomass and plant species composition will continue to fluctuate in response to weather irrespective of what is done to the herbivores. The upper and lower limits of plant biomass are imposed more by the weather than by grazing pressure.

(2) *Stabilise the herbivore populations*

Even though the biomass and composition of the vegetation cannot be stabilised in this climate, that of the herbivores can be. It requires decoupling the dynamics of the animals from those of the plants. First a desired density of herbivores is decided upon. Then it is enforced by culling animals when they threaten to exceed that density, and feeding them when pasture biomass drops below the level necessary to maintain that density. A less stringent objective allowing fluctuation within limits would be achieved by the same means.

(3) *Hold the herbivores below a prescribed limit of density*

Culling will be required when the limit is approached. Since no lower limit is identified, no action is required to arrest declines.

(4) *Hold the herbivores below carrying capacity*

This is still a commonly stated goal in some countries although less common than it was twenty years ago. In all such cases known to us the managers were in fact advocating a condition of the vegetation rather than a density of animals, their model being the pasture appropriate to animal production (see Section 11.23). If this were the stated objective, park managers would be well advised to re-examine the logic by which it was selected before deciding on the action by which it might be achieved.

(5) *Preserve selected plant associations*

If the plant associations selected for preservation cannot cope with the extremes of grazing pressure resulting from the system running itself, culling will be necessary to hold grazing pressure below the elimination threshold of the association most vulnerable to grazing. As Bell (1981) reminds us, the decision on which plant associations are to be favoured is a judgement of value, not a technical judgement. Walker & Goodman (1983) warn that such action may eliminate other plant associations that depend upon grazing for their continued existence.

(6) *Re-create the system as it was at some previous date*

We choose 1835 for purposes of illustration, the year in which the first European entered the area. We are not certain what the system looked like in those days but there is some evidence that kangaroos were less abundant than they are today. Thomas Mitchell travelled along the Darling from the site of present-day Bourke to Menindee in the winter of 1835. Although the party was accompanied by dogs that were used in other areas for hunting kangaroos, no kangaroo or emu was seen along the lower half of that stretch of the Darling (Mitchell, 1839, p. 308). In those days an additional grazing pressure was applied by smaller herbivorous marsupials, the brush-tailed bettong *Bettongia penicillata*, the burrowing bettong *B. lesueur*, the bridled nail-tail wallaby *Onychogalea fraenata* and the eastern hare-wallaby *Lagochestes leporides*, all now extinct in this region. We do not know whether the average plant offtake at that time was less than, equal to, or greater than it is on the park today.

Should these species be re-introduced to Kinchega as a first step in recreating the world of 1835, they would be entering a system different from what it was. It would be necessary to determine experimentally which conditions favour their re-establishment and which do not. Such experiments would best be carried out in several large enclosures, and the factors to be tested would include the grazing pressure of kangaroos. Control of kangaroo density would be required in at least one of those treatments.

(7) *A healthy population in a healthy habitat*

The reasoning behind this objective is that, in most systems, a reduced density of herbivores results in increased fecundity, decreased natural mortality, larger animals and healthier animals. It is a direct result of the lowered competition for food. Plant biomass is higher and the proportion of the ground shielded by plants is greater, a reflection of reduced grazing pressure. The objective is attained by continual culling, the more culling the more healthy the plants and remaining animals. It is often advanced as a corollary of objectives (1), (2) and (4) above rather than as a primary aim.

Attractive as the objective might appear it is not without problems. The very real increase in the health of the surviving animals, and the enhanced verdancy of the plants, may be achieved at the expense of ecological resilience. Walker (1981), Walker & Goodman (1983) and Caughley (1983) identified a number of mechanisms whereby artificial reduction of herbivores can erode the system's ecological defence mechanisms. 'Restraining the normal dynamics of a population or some community of populations, attempting to keep it constant near some defined (or supposed) equilibrium — often by culling — is usually counterproductive with respect to what is hoped will be achieved. It leads to a decline in the resilience of the system concerned, and it is therefore important to let the natural variation in numbers take place.' (Walker, 1981, p. 57).

(8) *Conserve ecological processes*

With this objective the aim is neither to conserve specific ecological states nor to dampen the dynamics of specific plant and animal populations. The system is encouraged to run itself. Changes in ecological states and fluctuations in animal and plant density are expected and accepted. Culling of herbivores and predators is inimical to this objective, as is feeding animals artificially or regulating their supply of drinking water.

11.3 Management of kangaroos on rangelands

In this section we explore the value judgements involved in the kangaroo debate and note the arguments for and against killing kangaroos (the words killing and culling are here used interchangeably.) As background we offer information on the effect of kangaroos on rangeland and on rangeland agriculture, the economics of the pastoral industry, the operation of the commercial kangaroo industry, and the constraints imposed on kangaroo management by quirks of the legal system. We then discuss three options for management of kangaroos on rangelands: full protection, culling for 'damage mitigation', and harvesting as a renewable resource.

11.31 *Background*

Some people favour killing of kangaroos on rangelands. Others are against it. Each of these positions is supported by several groups or organisations that have different motives and different views. All participants in the debate have access to the same basic information. Here is the common ground:

The main concentrations of kangaroos are on the sheep rangelands of the inland where they are probably now more abundant than when the sheep arrived. Numbers fluctuate widely in response to the rainfall of the preceeding few years: a run of good years leads to large increases and drought leads to large decreases.

Total numbers of the three main commercial species were estimated by aerial survey as 19 million in 1981. In that year 1.5 million were harvested commercially. It is not known how many were killed under damage mitigation permits and not sold, nor how many were killed illegally. Although the 1982-83 drought in eastern Australia reduced numbers considerably (J. Caughley, Bayliss & Giles, 1984; G. Caughley, Grigg & Smith, 1985) many millions remain.

These statements are as near to facts as one is likely to get. Seldom are they seriously debated. Yet there is no agreement on the right option for managing kangaroos. The three basic standpoints and their underlying motives are summarised in Table 11.1. Although such categorisation rides roughshod

Table 11.1. *A taxonomy of attitudes towards kangaroos.*

All shooting to stop
> (a) animal liberationists: kangaroos, together with other species of animals, deserve much the same consideration as we extend to members of our own species; to shoot them is immoral.
> (b) kangaroo protectionists: shooting kangaroos is undesirable. Often linked with kangaroos being a national symbol.
> (c) some conservationists: there is insufficient scientific evidence that kangaroos cause significant agricultural damage. Killing cannot be justified.

Shooting allowed for damage mitigation
> (a) pastoralists: kangaroos compete with livestock, at least at some times and in some places. They should therefore be controlled. Regulation should be minimal and commercialisation should be allowed.
> (b) some conservationists: where actual or potential damage can be demonstrated kangaroos may be controlled. Regulation should be tight. Commercialisation is a much debated issue.
> (c) some wildlife authorities: New South Wales National Parks and Wildlife Service, South Australian National Parks and Wildlife Service, Victorian Fisheries and Wildlife Service.

Shooting allowed for commercial gain
> (a) some resource ecologists: kangaroos can be treated as a renewable resource and harvested accordingly. Most require regulation of the harvest.
> (b) some pastoralists: probably a minority because kangaroo products may affect markets for livestock products. Minimal regulation is required.
> (c) kangaroo industry: wildlife authorities may choose to view the industry as a management tool but the industry is quite clear as to its priorities. Regulations that enhance the orderly flow of operations are welcomed.
> (d) Some wildlife authorities: Queensland National Parks & Wildlife Service and perhaps also the Western Australian Department of Fisheries and Wildlife. Regulation is considered necessary.

over the distinctions on which splinter groups are founded, this table may be a useful form guide for those uninitiated in the kangaroo debate.

Kangaroos are perceived as a pest of agriculture, an animal of variable commercial value, an object of conservation interest, and a cause célèbre for some groups within the animal welfare/animal liberation movement. All viewpoints except the last have been around for at least a century (Denny, undated [1982])

The history of commercial harvesting has been outlined by Livanes (1971), Denny (undated [1982]) and Prince (1984a, b). Kangaroos have been utilised commercially since the early days of the New South Wales Colony (Governor Macquarie put a landing tax on kangaroo skins in 1802), but major exploitation did not start until the 1850's. Trade was mostly in skins until refrigeration became economically feasible in the 1950's. Trade in whole carcases, as against skins alone, is now more common and is actually required by the regulations of some states. Pastoralists have killed kangaroos for profit, as pests, for sport, and to provide food for domestic carnivores. Their methods are generally less refined than those of the commercial hunters and include the use of snares, dogs, traps, poison, and drives (Gooding & Harrison 1955; Denny, undated [1982]). It is only within the last few decades that commercial harvesting and the interest of pastoralists in protecting agricultural production have been linked by kangaroo management programs.

Concern for the survival of the large kangaroos was first voiced as early as 1822, 34 years after the colony of New South Wales was founded. It was echoed by eminent naturalists such as Gould, Darwin, and Boldrewood during the remainder of the nineteenth century. Public opinion and the law took some time to catch up. New South Wales will serve as our example.

Marsupials were declared noxious under *Act No. 11.44. Vic. 1880* ('An Act to protect the pastures and live stock of the Colony from the depredations of certain noxious animals') and it was the duty of landowners and lessees to destroy them. In 1903 some protection was given to red kangaroos and wallaroos (*Macropus robustus*) under the *Native Animals Protection Act, 1903*, but this Act also recognised the commercial value of fauna. In 1918 protection was extended to many other marsupials (including other large kangaroos) by the *Birds and Animals Protection Act*, 1918-30. Although the legislation has been redrafted and amended several times since 1918 (it now is incorporated in the *National Parks and Wildlife Act*, 1974), its basic structure is unchanged.

The major conservation thrust did not come until the 1970's. In the early part of that decade there was heightened activity by conservation groups, a major symposium on kangaroos (*Australian Zoologist* 16(1), 1971), action by the Commonwealth (Federal) Government to ban exports of kangaroo products (1973), and action by the U.S Government to ban their import (1974). (Australia is a federal system consisting of a central government with restricted powers [Commonwealth Goverment], six semi-autonomous states, and two territories that are under Commonwealth control. The Commonwealth Government has power over export and import.) Elaborations and improvements to the kangaroo management plans of some of the states soon followed the bans, and the Commonwealth Government lifted its export ban in 1975. The U.S.A. did not lift its import ban until 1981. In that year new export rules were implemented by the Commonwealth Government under the Customs Act, 1902 (subsequently modified and incorporated in the *Wildlife Protection [Regulation of Exports and Imports] Act*, 1982), essentially to force state compliance with the 'national plan of management for kangaroos'. An inquiry into the management of kangaroos was begun in 1984 by the Select Committee on Animal Welfare of the Australian Senate.

The large kangaroos (red, eastern grey, western grey, and euro or wallaroo) are harvested commercially in only four states and in neither territory. Although these four states now have kangaroo management programs and would claim to comply with the national plan, there is substantial variation between states in the way they control the killing of kangaroos and in the justification they advance for permitting it. Their plans of management are set out in 'Kangaroo Management Programs of the Australian States' (Anon., 1984). Two of those states (New South Wales and South Australia) rely heavily on mitigation of damage to crops and fences, and competition with sheep for pasture, to justify the killing. Queensland's plan quite clearly rests on harvesting of kangaroos as a renewable resource; conflict with agriculture is tendered merely as a supporting reason. Western Australia's position is not entirely clear. Parts of their plan invoke damage mitigation and others advocate harvesting of a renewable resource. Although Victoria and the Northern Territory do not allow commercial harvesting of kangaroos, they allow licensed non-commercial shooting for damage mitigation with the number taken not being subject to quotas approved by the Commonwealth Government nor audited by a tagging procedure.

Publication of these programs has not diminished the public debate on management of kangaroos. Views recently presented to the Senate Select Committee on Animal Welfare were as polarised as ever. The published programs in fact made the various wildlife authorities clearer targets for the displeasure of many groups who claimed, sometimes justifiably, that objectives were not defined clearly or that the methods employed could not achieve the stated aims.

Before discussing three options for management we will canvass a few topics that have, or should have, a substantial influence on that choice. These can conveniently be stated as questions:

(i) What is known about the influence of kangaroos on rangeland vegetation?

(ii) How do kangaroos influence the practice and profitability of rangeland agriculture?

(iii) What is the structure and behaviour of the kangaroo harvesting industry?

(iv) What are the present legal constraints on the range of management options?

11.32 *Impact of kangaroos on rangelands*

Most authors believe that kangaroos have increased substantially in the eastern pastoral zone since European settlement (Frith, 1964; Newsome, 1975; Poole, 1978; but not Denny, undated [1982]). The position in Western Australia is less clear (Prince, 1984a, b). The putative increase in the east is ascribed to a proliferation of watering points, to modification of the vegetation by sheep, and to the virtual extermination of native dogs (dingos). Additionally there has been an enforced change in the lifestyle of aborigines who may previously have exerted a considerable pressure of predation on kangaroos around sparse sources of water. If the majority are correct, kangaroos in the eastern pastoral zone can be regarded as what range managers call 'increaser species'. Kangaroos thus may now be affecting the composition and biomass of vegetation and hence be an influence, along with sheep, on the habitat of other species, domestic and wild, more so than they were 120 years ago. That possibility has not yet been addressed by any long-term study.

The complaints levelled by pastoralists against kangaroos include that kangaroos compete with livestock for native pastures and water, damage crops at the early growing phase, eat irrigated and improved pastures, damage fences, and cause numerous road accidents. All that is true, but the principal source of unhappiness is competition with livestock for native pasture.

The relevant dietary studies have been canvassed in Chapter 5. Careful review of those studies indicates that there have been no quantitative and comparative studies of the diet of kangaroos and livestock that would withstand rigorous scientific scrutiny (Australian Senate 1985 pp. S 3024-49). However, there have been several qualitative studies, some comparative, that indicate a broad overlap of diets between kangaroos and livestock. Dietary overlap, of itself, is unlikely to precipitate competition in good seasons, first because resources are more than adequate for both kangaroos and livestock, and second because individual herbivore species are able to exert their dietary preferences. However, competition for preferred items (i.e. for grasses, forbs, and some chenopods) may be expected when food is scarce. Sheep can use some perennial shrubs (bluebushes and saltbushes) during droughts when there is nothing much else to eat. Although kangaroos will eat these shrubs, they do not survive on them indefinitely (Chapters 4 and 5). No comparative study of diet has run the full gamut of seasonal conditions in the pastoral zone. That partly explains our lack of understanding of when competition occurs and its magnitude, and our inability to assess the importance of competition to the overall functioning of a pastoral enterprise.

Kangaroos need affect the pastoral enterprise at only one point in a seasonal cycle before pastoralists perceive a need to reduce the numbers of kangaroos. In general pastoralists perceive kangaroos as pests. Sixty-eight percent of pastoralists surveyed in western New South Wales classed kangaroos as a major constraint on production (New South Wales Parliament 1983, 3rd Rep., p. 53).

Economics of pastoralism

The impression emerges from Chapter 10 that more livestock could be run on rangelands if the density of kangaroos were reduced. Australian rangelands are primarily used for low-intensity grazing. Sheep predominate in the south and cattle in the north. The few large towns (population exceeding 15,000) are associated with mining, and sparsely scattered smaller towns cater for pastoral activities and essential services such as roads and communications. Pastoral lands are generally held under a long-term lease from the state government, with all states now considering there is a need to control pastoral activities if rangelands are to be passed to succeeding generations in usable condition. This is usually approached by setting maximum carrying capacities (in terms of sheep equivalents) for each parcel of land. In New South Wales alone there are 31,000 km^2 of pastoral land held under such tenure.

Australian studies of the economic performance of pastoral enterprises are few. None weigh the private benefits of rangeland agriculture against the public cost. The most recent and comprehensive study of pastoral performance was conducted in New South Wales for the three financial years from 1977-78 to 1979-80 by a firm of agricultural consultants (Hassall & Associates, 1982). Their survey covered 10% of properties in the Western Division of New South Wales. Some of the data were reworked for the 'Inquiry into the Western Division of New South Wales' by a Joint Select Committee of the New South Wales Parliament (New South Wales Parliament, 1983). What follows is taken from the Second Report of that Committee.

The report shows that pastoral enterprises in the Western Division are performing poorly relative to other agricultural enterprises in New South Wales and that the trend is for this relative performance to worsen. Table 11.2 gives indicators of economic performance for the three financial years from 1977-78 to 1979-80. This shows that 46% of properties generated no disposable income and that financial difficulties were most common on properties with a low authorised stocking total (maximum carrying capacity expressed as sheep equivalents). Stocking above the authorised level was most common on properties rated below 6,000 sheep equivalents.

To avoid economic pressures on lessees that may lead to chronic overstocking, a minimum rated stocking for leases was canvassed by both Hassall & Associates and the Inquiry. Hassall & Associates suggested that a property size of 23,000 sheep equivalents would be required by 1985-86 to maintain a family. The Inquiry criticised this as unsound and opted for a much smaller increase to a minimum of 6,000 sheep equivalents. At 6,000 sheep equivalents

Table 11.2. *Indicators of pastoral performance in the Western Division of New South Wales for the financial years 1977–78 to 1979–80. Data[1] are arranged by property size (expressed as authorised flock size).*

| | Authorised Flock Size (Sheep equivalents x 10^3) | | | | | | |
	2–3	3–4	4–5	5–6	6–7	7–9	)9
Average farm income[2] ($)	10,901	19,406	27,116	34,475	29,266	44,773	72,113
Percent of farms with no surplus income[3]	61	46	58	30	35	28	25
Ratio of sheep equivalents carried to no. authorised	1.16	1.50	1.31	1.38	1.15	1.10	n.a.

[1] Data are from the Second Report of the Inquiry into the Western Division of New South Wales (New South Wales Parliament, 1983).

[2] Gross income minus operating costs = amount that farm has left to meet family expenses, repayment of principal, taxation, and investment in new technology (including servicing new debts). The Report estimated that an average of $22,799 was required for family expenses, repayment of principal, taxation and replacement of plant.

[3] Amount available for capital improvement, investment and non-essential items. Note that 30% of properties had average deficits greater than $10,000.

35% of lessees faced financial difficulties over the survey period, and a realistic minimum undoubtedly lies between the two.

Stock numbers in the Western Division of New South Wales have averaged 8.1 million sheep equivalents since 1950 (7.2 million sheep and 100,000 cattle; Hassall & Associates, 1982). Using the 8.1 million average, an increase in minimum lease size to 10,000 sheep equivalents would see the 1,490 properties of 1980 reduced to 810. The survey showed that leases with rated capacity of more than 9,000 sheep equivalents averaged a net farm income of $72,113 p.a. before tax, living expenses, and repayment of principal. Even though these properties performed relatively well, surplus annual income for capital improvements, debt servicing, and so on, was well below $40,000 per annum.

Would 810 such properties contribute significantly to the wealth of New South Wales? The figures suggest not. If all direct and indirect government subsidies were taken into account the balance sheet might well record a substantial net loss. Since that macro-economic analysis has never been attempted our suggestion remains only a speculation. For political and social reasons alone it is unlikely that abandoning rangeland agriculture in New South Wales would be seriously considered. However, the poor performance

Fig. 11.1. The merino is the universal breed of the sheep rangelands.

of many Western Division leases, and the resulting chronic overstocking noted by the Inquiry, emphasises the need to examine rangeland use holistically rather than simply to look for short-term solutions to falling farm income.

Kangaroo populations were at high density in New South Wales throughout the period of that economic survey, 8.2 million being estimated for the western plains in 1980 (J. Caughley & Bayliss, undated [1982]). However, there was substantial year-to-year variation in farm incomes over the three years of the economic survey, suggesting that although kangaroos might have exerted a background effect, they were not a major influence upon profitability over those years.

Although the Report noted that some properties of all sizes were in financial difficulty, the main factor determining profitability of a pastoral enterprise was its rated carrying capacity. Since kangaroos are presumed to reduce carrying capacity, the relationship between removing kangaroos and increasing carrying capacity is of interest. The long-term aspects were discussed in Chapter 10.

11.33   *Commercial utilisation of kangaroos*

The kangaroo industry comprises a large number of shooters (many of whom are part-time) and relatively few wholesaler/processors. The activities of both are controlled by the various state wildlife authorities. Controls include limiting the number of shooters, restricting their area of operation, and nominating the numbers of kangaroos that may be taken and the product that may be offered for sale (e.g. shooting for skins alone is restricted severely in New South Wales). The primary objective of the administrative control is to avoid over-exploitation when the market is bouyant. However, since some wildlife authorities regard the kangaroo industry as a management tool, the nursing of the industry through bad times becomes a secondary objective. Wildlife authorities attempt to achieve this by limiting the number of authorised buyers of meat and skins, thereby limiting the competition between them. Profits in good years may then buffer the lower earnings of bad years. An alternative would be to allow the industry a maintenance quota irrespective of population trends, but this would alter the status of the industry from a management tool to a legitimate enterprise in its own right.

Wholesalers will buy kangaroos from shooters only if they can make a profit from selling the products. The number killed therefore depends on availability of markets for meat and skins.

Skins are used in the fur and leather trades. Close substitutes exist (calf and kid) and when these are readily available prices for kangaroo skins fall to low levels. Prices paid to skin shooters therefore often fluctuate sharply (e.g. in 1979-80 prices varied from $1.50 to $7.00 per skin), and vast stockpiling of unsold skins may occur (Shepherd & Giles, undated [1982]).

Kangaroo meat is used both for pet food and as meat for human consumption. Meat used for pet food is primarily sold on the domestic market, mainly in Sydney and Melbourne, with a small amount being exported from Australia. The domestic pet food market takes meat on two criteria — price and acceptability. Beef and sheep meats are more 'acceptable' to the public and will compete successfully when their price approaches that of kangaroo meat. As a result, the size of the kangaroo meat market will vary with the frequent variations in the fortunes of the sheep and cattle industries.

The Australian pet food markets cannot currently cope with the potential supply of kangaroo meat from both New South Wales and Queensland and therefore demand usually limits the harvest. Supply is occasionally limiting: for example, markets are sometimes buoyant when kangaroos are difficult to obtain because of decreased abundance or because of reduced accessibility for shooters during times of widespread rain.

Fig. 11.2. Kangaroos within the rangelands are shot by professional shooters for meat and skins. Carcasses bear individually numbered tags required by the state wildlife authority which oversees the harvest.

Meat for human consumption is primarily exported as game meat. Between about 1955 and 1969 there was a substantial export trade in game meat (MacFarlane, 1971) that was closed down by the importing countries for a number of reasons, including poor meat quality, high salmonella contamination, contamination by vegetation and dirt, and infestation by the parasite *Dirofilaria roemeri*. The export trade resumed in 1980 under supervision of the Commonwealth Department of Primary Industry. There is little doubt that importing countries will reimpose restrictions if meat quality fails to meet health standards.

A small domestic market for kangaroo meat for human consumption was opened in South Australia in 1980. The possibility of a similar market in New South Wales is remote, and the situation in the other mainland states is unclear.

Market forces operate independently of damage mitigation needs and are often out of phase with them. Table 11.3 shows this in the form of a comparison between authorised quotas and the actual commercial cull in New South Wales for the period 1975 to 1984. The carcase industry in particular is unable to contract or expand rapidly to meet short-term fluctuations in perceived damage mitigation needs; it would also have difficulty with the longer term fluctuations predicted in Chapter 10. Table 11.3 also emphasises the reality of the separation between the goals of the industry (to make money) and the perceived goals of the wildlife authority (to control populations).

Thus the industry cannot be a reliable partner in a kangaroo management

Table 11.3. *Authorised and actual commercial harvest of kangaroos (reds, eastern greys and western greys combined) in New South Wales from 1975 to 1984.*

Year	Authorised harvest	Actual harvest
1975	212,900	123,000
1976	319,400	96,700
1977	321,500	167,200
1978	345,000	220,000
1979	645,000	520,000
1980	645,000	677,534
1981	694,500	488,647
1982	843,000	664,342
1983	843,000	400,477
1984	500,000	229,484

program unless the quota rather than the market becomes the limiting factor and there are efficient controls that ensure the quota is not exceeded.

11.34   *Legal constraints*

In this section we outline some legal matters that must be considered when developing a wildlife management program in Australia.

Any such program will have to be effected through the legal system and, at some stage, a lawyer will have to convert biological objectives into law. Legislation is difficult to amend. In most instances it is also sufficiently precise to preclude its effective use to achieve objectives other than the original. It follows that a clear definition of objectives, and methods for achieving those objectives (including thorough investigation of enforcement procedures), should precede legal drafting. The drafting phase requires both biologist and lawyer to overcome the language barriers imposed by their respective disciplines.

It has already been noted above and elsewhere (Shepherd & Giles, undated [1982] and Australian Senate 1984 pp. S2068-75) that the existing kangaroo management programs are not fully effective. One of the main reasons is that the legal framework is often at odds with a shifting set of wildlife management objectives. Perhaps the lesson is that wildlife managers should not enshrine muddled thinking in law.

When framing programs, legal constraints that may limit the actions of a wildlife authority must be considered. These may be higher in the legal hierarchy than the wildlife law, be outside the jurisdiction of the legal system enacting the wildlife law, be generally applied legal principles, or be existing statutes and policies in related fields. These constraints do not prevent the Australian states from legislating to control the killing of wildlife. If a culling program simply required that specified animals (or a class of animals) be destroyed under specified conditions there would be no problems except for the logistics of enforcement arising from the need to have a ranger behind every tree. However, in most states a commercial harvest is part of the kangaroo management program, and in the search for a method to enforce the provisions of that program the state government usually attempts to regulate this enterprise. At this point the constraints start to bite.

For state wildlife authorities the main problems lie in the hierarchical constraints. These are imposed either directly by the Constitution (*Commonwealth of Australia Constitution Act*, 1900 [U.K.]) or by valid legislation pursuant to a head of power granted to the Commonwealth by the Constitution. The former is exemplified by Section 92, which effectively prevents wildlife authorities from controlling interstate movement of fauna or their products.

The latter becomes a problem because the Commonwealth Government likes to dabble in the wildlife area. It has no direct head of power relating to wildlife, but gains partial control by using other heads of power (for example, control over export of kangaroo products enables some control over management programs that depend on commercial harvest of kangaroos). Because indirect use of Constitutional powers is subject to changing interpretations of the High Court, the validity of Commonwealth 'initiatives' is always open to doubt.

To be effective wildlife law must be enforceable. The chance of detecting a breach must be high, or, if low, the penalty must be high enough to make the activity a risky proposition. In either case the courts must support the wildlife legislation fully. In general they have not done so (Shepherd, 1981).

11.35   *Options for management*

We now consider three options for managing kangaroos: full protection, culling for damage mitigation, and harvesting as a renewable resource. Each of these is constrained by problems of enforcement and by their acceptability to the community.

Enforcement must be aimed at the act of killing and subsequent possession. In the absence of a commercial harvest system this is exceptionally difficult. The sheep rangelands are vast and few wildlife rangers are stationed there. Most kangaroos are shot at night when most rangers are at home. A commercial system theoretically makes enforcement easier because it centralises some activities at definable points. However, it also provides an incentive to evade the law for profit.

11.36   *The option of full protection*

There are two circumstances in which the full protection of kangaroos might be appropriate. The first is where any hunting is judged to endanger one or more kangaroo species. The other is where the animal's intrinsic worth is perceived to be such that hunting them is aesthetically repugnant.

Most protection laws are passed to conserve a species, and in fact the partial protection conferred on these three species by all states reflects that motive. Unrestricted hunting, it is judged rightly or wrongly, could endanger the species. Since there is no indication that the present level of hunting of kangaroos on the sheep rangelands is causing a long-term decline, and good evidence that it is not, there is no scientific justification on grounds of conservation for additional protection up to and including full protection.

The aesthetic motive for full protection has two heads. The first is 'animal liberation'. The essence of 'animal liberation' is that regard for an individual animal is independent of, or almost independent of, the species to which it belongs, and independent of its conservation status and of whether it is wild or domesticated. We consider this philosophy is unlikely to gain wide support in the forseeable future and will thus not consider it further in the context of protecting kangaroos.

The second aesthetic motive for full protection is more widespread. 'Charity' will serve as its label. We may cherish a species for its grace, beauty, gentleness or scientific interest, and so greet the death of one of its individuals as a personal loss. The 'charity' motive in Australia has gained ground over the last decade. In part that reflects the kangaroo being the national symbol — there is something not quite decent about shooting national symbols — and in part it reflects a widespread misconception that kangaroos are endangered. But when that misunderstanding is excised the charity remains.

Full protection would be difficult to enforce. That is not necessarily a telling argument against it, but the consequences of failure should be considered. The main one is that pastoralists would continue to kill kangaroos and they would do it with less finesse or humanity than commercial harvesters. This would be compounded by local rural community values. These are in tune with kangaroo shooting and, unless cruelty were involved, a defendant could expect a sympathetic hearing on any charges of killing or possession pursued in court.

There may be ways of softening the impact on landholders of a total ban. Compensation seems attractive in that the 'people' are paying to have 'their symbol' survive on privately held land. The simplest approach would be annual population assessments with payment according to density.   Alternative procedures would be required where special damage was claimed (e.g. isolated crops on pastoral land), or in areas where aerial survey was not feasible. The real cost of kangaroos to primary production would have to be estimated before such a scheme could be designed rationally.

We can estimate the likely long-term effect of full protection upon kangaroo numbers. Where present offtake is close to the long-term maximum yield, the kangaroos are being held at about 40% below their long-term unharvested mean density (Fig. 10.5). Where offtake is lower, the disparity between harvested and unharvested density is proportionately smaller. An exception is where they are overharvested in the sense that they have been harvested down to a level below that generating long-term maximum yield. We suspect that is uncommon on rangeland but may be quite common in parts of the sheep-grain belt.

11.37   *Culling for damage mitigation*

Kangaroos can cause damage to fences and crops, but the major charge against them on rangeland is that they utilise pasture in much the same way as do sheep. It is the competition with sheep, potential or actual, that lies behind most applications for a 'damage mitigation' permit. In its simplest form this option should provide most things to most people: kangaroos cause hardship, that is verified during an inspection by a ranger, and a permit is issued for the destruction of a specified number of a particular species. There is no basic problem with the objective or its translation into law.

However, logic requires a nexus between the hardship and the number on the permit. Further, the 'damage' assessed should be that anticipated over some future period; a permit relating to damage already caused is simply punishing kangaroos for wrongdoing. Anticipated damage can only be past damage extrapolated and, as Chapter 4 shows, the change in conditions from those in which kangaroos and sheep might be expected to compete to those in which they do not compete may be sudden and unpredictable. There is also the problem of an objective field damage assessment technique — there isn't one.

Culling should be species-specific and numbers killed should be accurately controlled; shooting therefore becomes the best method available. Shooting could be carried out by the pastoralist, commercial hunters, recreational hunters, or the state. Evidence given to the Senate Inquiry suggests that state-funded shooters are not a proposition because of cost (Australian Senate 1984 p. 1872), and recreational hunters are unlikely to have much effect on numbers because most do not class kangaroos as game animals. This leaves only pastoralists and commercial hunters (which can be the same person if the pastoralist involves himself directly in commercial hunting). The decline in employment of farm labour over the last twenty years (New South Wales Parliament, 1983, Second Report, p. 79) ensures that pastoralists are usually too busy with other tasks to put much time into shooting kangaroos. Commercial culling was discussed in Section 11.33 above, with the conclusion that a successful kangaroo management program could not depend upon commercial culling alone unless markets ceased to be limiting and the industry developed a capacity to expand and contract rapidly.

The problem of removal is exacerbated because in any given locality the agricultural production pattern is fairly uniform and therefore any kangaroo damage of economic significance is likely to occur on many properties over a short period. Further, the wildlife authority needs to receive applications, carry out inspections, issue permits, and enforce regulations. A few commercial shooters and wildlife officers in a region cannot hope to make such a system work according to the rules.

If it is not feasible to assess damage on individual properties, then culling for damage mitigation will have to be put on a different footing. We are still obliged by logic to consider potential damage and we must therefore have some idea of when kangaroos are likely to come into conflict with pastoral enterprises. Chapters 5, 6, and 10 together suggest that, in the study area, competition may commence when biomass of available vegetation falls below 300 kg/ha and that its intensity will be a function of the combined densities of the sheep and kangaroos. Biomass is likely to fall below this level quite often.

There are a number of strategies for suppressing kangaroo density. These range from intermittent culling when the population is at high density to continuous harvesting of a fixed proportion of the population. Some strategies are undoubtedly more efficient than others. The important point is that a clear objective is required within the suppression umbrella before a strategy is selected. A primary constraint is that we should avoid suppressing density to the point where kangaroo populations are threatened by a run of poor seasons.

In our fluctuating environment, some strategies for suppression of density will require monitoring of changes in rate of increase. Rate of increase is estimated from two indices of density, usually obtained by aerial survey. This technique is essentially restricted to winter and has a substantial detection lag for increasing populations (Chapter 8) but none for decreasing populations. However, there is a close relationship between rainfall and rate of increase in the rangelands (Chapter 8; J. Caughley, Bayliss & Giles, 1984) and so it should not prove too difficult to predict population responses and to check those predictions by an annual aerial survey.

Whatever strategy is chosen there will be the problem of who removes the kangaroos. The system of administering offtake will also be quite different from that appropriate to individual properties. For most strategies quotas will be required for specified areas, their size being determined by the uniformity of conditions and by the precision of aerial survey results as influenced by the size of each block surveyed. As a rough approximation, the coefficient of variation of the estimate of density is increased by 40% when the size of the block is halved. Although under this scheme the unit of assessment would no longer be the individual property, it may still be desirable to spread culling throughout the designated area by allocating permits to properties or by some other strategy.

11.38   *The option of harvesting a renewable resource*

In contrast to culling for damage mitigation, the primary aim of resource harvesting is to take and utilise a continuing yield. Harvesting thus has a right to exist independent of the need to protect agriculture. It may be achieved either by farming kangaroos or by harvesting them as wild animals.

Farming of kangaroos

Technical aspects that would limit our ability to farm kangaroos on rangeland include mobility, handling difficulties, fencing, relatively slow rates of growth and reproduction, and the absence of a reliable technique for censusing populations on small areas (Shepherd, 1983). Administrative difficulties are equally forbidding.

Landholders at present have no proprietary rights over kangaroos on their land. But for kangaroo farming, the pastoralist will require an interest sufficient to ensure that his investment in raising the animals cannot be lost to a third party without adequate compensation. A workable legal solution would be possible for intensive farming, but rangeland farming poses substantial legal problems caused by the mobility of kangaroos and the difficulty of identifying farmed animals. Captive kangaroos can be branded in some way and small areas can be fenced to prevent straying, but kangaroo-proof fencing is not economic on rangelands and mustering of kangaroos for branding is not feasible.

A partial solution to the mobility problem is to enlarge the kangaroo farming enterprise to increase its area to boundary ratio, either by enlarging a single property or by an agreement between adjoining landholders to run kangaroos as a joint enterprise. Enlarging a single property to reduce the problem posed by mobility requires two major changes to the existing system of land allocation. First, areas added to the property must be adjoining, whereas dispersed units are often amalgamated for sheep and cattle enterprises. Second, the maximum area allowed to each landholder would need to be increased (most states have restrictions on size of rangeland leases).

All these problems would be solved rapidly if the farming of kangaroos could provide an income double that accruing from sheep. With low prices and restricted markets there is no present possibility of that. Kangaroo farming is not now an economic proposition but its feasibility should be reassessed periodically.

Harvesting wild kangaroos

The offtake of a harvesting industry may be regulated by the size of the market, or by the population density at which cost of harvesting equals the reward from harvesting, but it is not regulated by the need to conserve the resource. The last point may seem illogical, but it has a secure rest in economic theory. The main reason is that the discount rate (the factor by which future earnings from harvesting must be discounted to estimate their present value) places greater stress on current yield than on future yield. Overharvesting can in many circumstances be economically sensible at the same time as it is ecologically disastrous. The theory is spelled out by Clark (1976) and summarised by May (1976), and its practice is demonstrated most vividly by the history of the whaling industry (Gaskin, 1982). The interested reader is referred to those publications for a full exposition. Here we simply note the principle and draw the corollary that the harvesting of a renewable resource, if it is to continue indefinitely, requires the imposition of a control mechanism that is totally independent of the economics of the industry harvesting that resource and only mildly sympathetic to its short-term economic problems. The responsibility for such control is currently held by the fauna authority of each state.

The history of regulation of the use of resources indicates that the regulating authority can very easily become locked into a symbiotic relationship with the industry to the detriment of the resource. Caughley (1977) gave an example from shore-based whaling in Western Australia during the early 1960's. The quotas set by the Commonwealth Government, far from being estimates of sustained yield, appeared to be predictions of the yields the industry would achieve if working at full capacity. In reference to the whaling industry's penchant for driving stocks to economic extinction rather than taking a sustained yield, Gaskin (1982) remarks that 'there is also evidence that in some countries this attitude by industry was actively or passively supported by those very government agencies naively believed by the general public to be deeply committed to the preservation or conservation of national and international renewable resources.' Here we simply flag this as a potential problem which could arise should markets expand and prices for kangaroo products escalate.

By the calculations summarised in Fig. 10.4 the maximum long-term yield from red kangaroos will accrue when between 10% and 15% of the population is harvested each year. That for the two species of grey kangaroo is likely to be lower, between about 8% and 12%. Such harvesting will reduce average long-term density by about 40% (Fig. 10.5). Sustained harvesting above those rates will reduce density further, resulting in a lower long-term offtake achieved at higher cost and by greater effort. Each of these conclusions holds only if the lowered density of the kangaroos occasioned by the harvesting is not compensated for by an increase of sheep.

Harvesting systems

Kangaroos may be harvested according to an annual quota that remains fixed (either as a proportion of the population or as a fixed number of animals) irrespective of fluctuations in density, or according to a quota adjusted each year to track density, or under a system holding hunting effort constant, or under a system where hunting effort is varied in response to changing density. Each of these systems has its good points and its hazards, both administratively and biologically.

Harvesting a constant number of animals each year is inefficient when applied to a population subject to large, environmentally induced, swings in density because the quota must be set low enough to be safe at the lowest anticipated density. Its advantage lies in a reduced need for monitoring and a steady production that stabilises markets and labour requirements. A varying annual quota increases administrative flexibility, increases the long-term accumulated yield, but requires monitoring of the population at frequent intervals. That option may translate to a quota specifying a fixed proportion of the population. This has close conceptual affinity with fixed-effort harvesting but, for reasons stated in the next paragraph, is not as safe.

Regulation by quota can be dangerous when the quota is set close to the maximum sustained yield. The density at equilibrium with that yield is unstable (Beddington & May, 1977; May *et al.*, 1978) such that a small environmental perturbation may trigger a density slide. However if yield is controlled indirectly by limiting harvesting effort (in this context the number of hunters licensed), but with no restriction on the yield per unit of effort, that dangerous source of instability is eliminated. A fixed-effort system would, within limits, harvest the same percentage of the population at high and low density. Yield tracks density, the system automatically producing a higher yield when the animals are abundant and a lower yield when they are scarce. A regulatory mechanism is built into the harvesting system itself and it is thus reasonably safe providing that the appropriate harvesting effort has been calculated correctly. Further, its successful operation is less dependent upon a close monitoring of the population. That system could be refined for kangaroos by a slight downward adjustment of effort during droughts when the kangaroos are particularly easy to harvest. At this time the percentage taken per unit of effort is likely to increase.

The major disadvantage of regulating effort instead of yield is conceptual rather than evidential. Administrators hate it. They see it as untidy because the size of the yield is not totally under their control. An intermediate step has been inserted between prescription and outcome. The idea of a harvester being allowed to take a yield as large as his effort permits grates against the instinct to control and regulate. Nonetheless, fixed effort harvesting, as against fixed or variable quotas, has large advantages, particularly in a fluctuating environment. We do not know of any population being driven to economic extinction under a system of controlled harvesting effort. We can think of several dozen that suffered that fate under a system of open access or of quota regulation. The examples come from fish, whales, crustaceans and molluscs.

The pastoralist would not benefit directly from the harvesting unless he participated as a harvester or charged the hunter for the privilege of shooting on his land. On current profitability harvesters could not afford a substantial additional payment for shooting rights.

11.4   Conclusions

The arid-zone ecosystem analysed by this study is driven by water, but rainfall has scant seasonality and varies widely from month to month and from year to year. The behaviour of the system appears chaotic because it never stands still for more than a few weeks, but in fact it is governed by a tight chain of causality from rainfall, through growth rate of pasture, to rate of increase and decrease of the kangaroos. Although kangaroo density and pasture biomass vary greatly among years they constitute a robust centripetal system.

Periodic culling of kangaroos is not an ecological necessity in national parks. The system is quite capable of managing itself. However, different national parks have different goals, and culling may be necessary to attain an arbitarily chosen objective such as maintaining a plant community that cannot persist in the presence of heavy grazing. When such a decision is taken it should be identified clearly as a personal or departmental preference, not as an ecological imperative.

Culling of kangaroos on rangelands has a different function — to reduce competition with sheep. At a culling rate generating maximum long-term yield it reduces kangaroo density by about 40%. Issuing permits for damage mitigation on the basis of individual property inspection has some problems. In the absence of any objective criteria to assess either damage or local population size, permit issue becomes a social interaction — a wildlife ranger and a pastoralist trying to reach consensus on how many kangaroos should be removed. There is nothing rational about it from a wildlife management perspective, except perhaps that kangaroos should not theoretically become endangered under such a scheme.

Suppression of kangaroo density over areas larger than the individual property fits a little better with our present state of knowledge. Kangaroos and pastoral production conflict in the study area at pasture biomass levels below 300 kg/ha and that conflict is intermittent. Suppression of density is a form of insurance and insurance is designed for just such intermittent events. The kangaroo industry cannot be a reliable partner in such a scheme while markets for kangaroo meat and skins remain limiting. It cannot take enough animals to prevent a build-up of numbers occasioned by a run of good seasons. This often leaves a gap between the required offtake and the actual offtake.

No matter how kangaroos are removed, costs of removal, monitoring, and enforcement should be deducted from any profits made from the kangaroos, and the net figure should be balanced against the cost to pastoral production of having the kangaroos remain. This latter figure cannot be arrived at by assessing costs as perceived by pastoralists. It must be based on empirical data as to the real impact of kangaroos on the whole pastoral system. It would be difficult to justify marginally improving a pastoral industry by removing kangaroos at public expense if that pastoral industry were already in debt to the public purse overall.

Finally, we canvassed two further options for management of kangaroos on rangeland: full protection, and harvesting as a renewable resource. Full protection cannot be justified for reasons of conservation but it may be justifiable on aesthetic grounds. The decision is a judgement of value.

Little would be gained economically by farming kangaroos at the present time. Costs are likely to exceed profits. Harvesting of wild kangaroos is entirely feasible, need not be terminated over periods when the populations are in decline consequent on dry weather, and would best be managed by a regulatory system that controlled offtake indirectly by controlling harvesting effort. However, allowing exploitation of kangaroos purely for profit is also a judgement of value. Thus these conclusions on feasibility are not pronouncements on the desirability of such exploitation; that is an issue for another forum. If such exploitation did proceed, administrative costs should be borne by the industry. The greatest long-term yield from red kangaroos is taken at a rate of between 10% and 15% of the population each year. It is a little lower for grey kangaroos.

APPENDICES TO CHAPTER 8

APPENDIX 8.1

Correction factors for visibility bias of kangaroos counted on standard surveys from a light aircraft
Peter Bayliss

Vegetation cover

Index-manipulation-index (Caughley, 1977) was used to estimate total population size (and hence visibility bias) of kangaroos on Kinchega National Park. Rather than experimentally reducing the population by a known number of animals and comparing density indices before and after, their decline in numbers was estimated during a drought by counting carcasses on ground transects (Robertson, 1986). Most kangaroo deaths occurred in the four-month period from November 1982 to February 1983. Indices were obtained by aerial survey (see Appendix 8.2). The September 1982 and June 1983 indices were chosen for this exercise because these surveys were close to standard (i.e. clear skies and survey temperatures $\langle 20°C$). However there may still be biases because kangaroos altered their grazing behaviour when pasture biomass decreased during the drought: they spent more time grazing later in the day and hence were more visible.

Red kangaroos
September 1982 index = 17.8/km²
June 1983 index = 11.4/km²
Carcass count February 1983
= 11.7/km²
17.8 (*CF*) - 11.4 (*CF*) = 11.7
Correction factor = *CF* = 1.83

Western grey kangaroos
September 1982 index = 11.2/km²
June 1983 index = 2.8/km²
Carcass count February 1983
= 21.3/km²
11.2 (*CF*) - 2.8 (*CF*) = 21.3
CF = 2.53

A comparison of factors estimated to correct for visibility bias of red and western grey kangaroos seen from the air on standard surveys of open habitat, obtained by other methods, is presented in Table A8.1. The above method was chosen because the counts of dead kangaroos provides direct control.

All correction factors except that of Caughley *et al.* (1976) apply to an environment containing 60% open shrubland and 40% open woodland, with the following aircraft survey variables: altitude 300 feet (91 m); transect width 100 m per observer; aircraft speed 90 knots (167 km/hr). The Caughley *et al.* (1976) correction factor refers to a survey over open grassland with widely scattered trees, but with the same survey variables as the others.

Table A8.1. *Estimates of correction factors.*

Method	Red	W. grey
1. Indirect experiment (Caughley *et al.*, 1976).	1.81	—
2. Indirect helicopter experiment, indices standardised to 15°C (Bayliss & Giles, 1985)[a].	1.84	2.30
3. Comparison of air and ground counts: air counts of Bayliss, June 1983 (see Table A8.2); ground counts of Short & Bayliss (1985)[c].	1.02[b]	2.88
4. Direct index-manipulation-index using drought mortality.	1.83	2.53

[a]In this experiment aircraft speed and width of transect per observer were kept constant at 130 km/h and 100 m respectively. The surveys were flown at three heights (15, 46, and 91 m) and at ambient temperatures ranging between 0°C and 38°C. A relationship was obtained between numbers seen on one hand and survey height and temperature on the other. As counts from a helicopter and fixed-wing aircraft at a survey height of 91 m (300 ft) were not significantly different (unpublished data), this relationship was used to give an indirect estimate of absolute densities by extrapolating to height of zero and survey temperature of 15°C. However the resulting correction factors are likely to be underestimated because the effects of strip width and aircraft speed (see Bayliss & Giles, 1985) are not included. Other biases may result from extrapolating a regression back to the *y*-intercept outside the range of data: curvilinearity in the relationship between height and numbers seen is a possibility that could not be tested (see Bayliss & Giles, 1985).

[b]This appears to be the odd one out but the standard error of this estimate overlaps the estimates obtained by the other methods.

[c]Short & Bayliss (1985) estimated aerial survey correction factors directly by using ground counts as estimates of absolute density. The aerial counts were obtained at the survey variable levels used on standard surveys in Australia (i.e. 76 m height, 200 m width of transect per observer and 185 km/h). Their correction factors are applicable only to surveys operating at those levels. However, at the time of their experiment Bayliss conducted similar surveys using variables pertaining to Kinchega surveys (91 m height, 100 m width of transect per observer and aircraft speed of 167 km/h). The resulting indices are presented in Table A8.2. These figures are specific to the region of the Park covered by the ground counts.

There was no significant difference in counts between time-of-day (AM/PM) for either species so the indices were averaged over consecutive surveys days and time-of-day. Table A8.3 is a summary of the densities estimated by aerial survey (Kinchega variables) and ground survey.

Table A8.2. *The aerial counts used in estimating correction factors by method 3 in Table A8.1.*

Date	Time of Day			
		AM		PM
1983	Reds	W. greys	Reds	W. greys
21 June	381	75	345	53
22 June	285	23	417	80
23 June	477	53	422	89
24 June	472	64	505	94
Mean	403	54	417	79

Table A8.3. *Comparison of the kangaroo densities obtained by air and ground surveys that were used to calculate correction factors by method 3 in Table A8.1.*

Species	Aerial Survey*		Ground Survey
	Index	Density (km²)	Density (km²)
Reds	410	16.7	17.0
W. greys	67	2.7	7.8

*The aerial survey covered 24.6 km².

Temperature and cloud cover effects

Temperature

Bayliss & Giles (1985) showed that the temperature of the survey (as measured after lift-off) affected the visibility of red and western grey kangaroos. They demonstrated that the effect was weak below a threshold ambient temperature of 20°C. Hence all indices are corrected by the formulae below if the survey temperature was greater than 20°C:

red kangaroos: $R = 0.035 + 0.030 \ (TS) - 0.0021 \ (TS^2)$
western grey kangaroos: $R = 0.230 + 0.025 \ (TS) - 0.0026 \ (TS^2)$

where R is the residual variance about the linear regression of log-transformed index of abundance on time between 1972 and 1979. Thus the analysis of the effect of temperature on visibility has been corrected for the continuing increase in density that held through the 1970s. TS is the temperature of survey. The correction factor (CF), to standardise indices to a survey temperature of 15°C (typical winter surveys) is: $CF = e^R$.

Cloud

Short & Bayliss (1985) showed that cloud cover reduced the counts of red kangaroos and increased those of western grey kangaroos. Their survey variables were a speed of 100 knots (185 km/hr), a transect width per observer of 200 m, and an altitude of 250 feet (76 m). Bayliss (unpublished data) conducted a similar experiment using the survey variables associated with this study and obtained similar results. With 7/8 and 8/8 cloud cover the counts of red kangaroos decreased by 44% and hence an appropriate correction factor to standardise to clear skies is 1.78. Western grey counts on the other hand increased by 49% and hence an appropriate correction factor to standardise to clear skies is 0.67. There appears to be a strong interaction between the effects of temperature and cloud cover on visibility. Air temperature can remain high even on overcast days, but its effect on visibility is confounded by that of cloud cover to an unknown extent. To avoid possible gross inaccuracies by correcting for both when temperatures were greater than 20°C on heavily overcast days, only one of the corrections was employed.

APPENDIX 8.2

Indices of abundance of red and western grey kangaroos on Kinchega National Park and surrounding properties between 1973 and 1984, and for Tandou between 1978 and 1984.

Year	Month	KINCHEGA Reds	KINCHEGA W.Greys	SURROUNDING PROPERTIES Reds	SURROUNDING PROPERTIES W.Greys	TANDOU Reds	TANDOU W.Greys
1973	Jul	267(214)[s]	112(90)[s]	79(63)[s]	19(15)[s]		
		326(261)[s]	140(112)[s]				
	Oct	333(266)[s]	96(77)[s]	42(56)[t]	14(19)[t]		
		144(115)[s]	26(21)[s]	(45)[s]	(15)[s]		
1974	Mar	159(206)[t]	43(64)[t]	74(135)[t]	20(48)[t]		
		145(200)[t]	60(97)[t]				
	Jun	145	100	95	28		
		138	104				
	Sep	340	120	119	37		
		413	157				
1975	Jan	463	173	143	49		
		531	146				
	Apr	305	88	97	28		
		331	112				
	Jul	329	117	96	58		
		329	180				
	Oct	575	202	141	27		
		529	182				
1976	Feb	227(385)[t]	54(117)[t]	115(149)[t]	32(47)[t]		
		318(439)[t]	81(131)[t]				
	May	532	119	147	49		
		449	166				
	Aug	769	213	390	151		
		852	229				
	Dec	526	161	165	44		
		332	95				
1977	Apr	565	216	166	100		
		742	262	—	—		
	Jul	466	173	302	96		
		479	224				
	Sep	494	240	355	134		
		620	214				
1978	Jan	440	143	190	77		
		531	163				
	May	858	256	621	116		
		751	283				
	Aug	599	207	230	43	177	86
		442	158		—	—	

		KINCHEGA		SURROUNDING PROPERTIES		TANDOU	
Year	Month	Reds	W.Greys	Reds	W.Greys	Reds	W.Greys
	Nov	298(705)[t]	69(236)[t]	155(403)[t]	40(156)[t]	105	85
		824	153			—	—
1979	Feb	529	167	131(170)[t]	45(96)[t]	—	—
		660	172				
	Aug	912	268	375	66	361	118
		834	257				
	Nov	703	185	200(356)[t]	48(33)[t]	251	92
		955	183			—	—
1980	Mar	402(493)[t]	113(154)[t]	226(356)[t]	47(92)[t]	180(265)[t]	49(87)[t]
		453(556)[t]	85(142)[t]			189	34
	Jul	669	279	446	172	170(303)[c]	129(86)[c]
		953	292			206	93
	Oct	796	360	372	126	318	171
		808	343			360	102
	Dec	483(860)[c]	315(211)[c]	219(269)[t]	110(150)[t]	335	107
		975	282			248	119
1981	Mar	700	237	314	62	306(422)[t]	69(111)[t]
		627(770)[t]	158(215)[t]			285	124
	Jun	418*	637*	305	185	258(459)[c]	395(265)[c]
		851*	772*			270	279
	Sep	979	319	137(178)[t]	40(59)[t]	326	185
		721(885)[t]	201(274)[t]			282	101
	Dec	757(929)[t]	230(313)[t]	316(436)[t]	164(265)[t]	417	199
		1112*	297			329(404)[t]	89(121)[t]
1982	Mar	755(927)[t]	213(290)[t]	312	163	350	213
		731(1008)[t]	221(356)[t]			271(482)[c]	200(134)[c]
	Jun	517(920)[c]	388(298)[c]	317	110	265(472)[c]	303(203)[c]
		748	323			269	173
	Sep	710	352	113	91	113(201)[c]	95(64)[c]
		675	525			349	189
	Dec	240(427)[c]	223(149)[c]	162(255)[t]	59(115)[t]	233(321)[t]	131(211)[t]
		272(484)[c]	233(156)[c]			211	148
1983	Mar	490	114	89(151)[t]	102(220)[t]	121(215)[c]	144(96)[c]
		292(520)[c]	147(98)[c]			90(160)[c]	105(60)[c]
	Jun	497	117	174(309)[c]	123(82)[c]	70(125)[c]	139(93)[c]
		389	99			93(166)[c]	132(88)[c]
	Sep	250(445)[c]	173(116)[c]	121	26	132	64
		323	86			106(189)[c]	81(54)[c]
	Dec	271(459)[t]	46(99)[t]	41(106)[t]	6(23)[t]	130(179)[t]	45(73)[t]
		272(334)[t]	84(115)[t]			61(75)[t]	45(61)[t]
1984	Mar	303	54	61	25	70	45
		248	48			47	37
	Jul	386	121	44	50	80(142)[c]	96(64)[c]
		290	81			86	60

The indices of abundance for each species are derived by summing counts of both observers over all transects for each survey (—— equals missing samples, and * equals inconsistent counts excluded from all analysis). An area of 39.9 km² was surveyed on the Park, 39.0 km² on properties abutting the park, and 38.5 km² on Tandou. In brackets are values of indices either corrected for the effects of temperature (^{t}), cloud (^{c}), or strip width (^{s}). The 1973 surveys using a 200 m / observer strip width are here corrected down to a 100 m / observer strip width by multiplying by 0.8 (Bayliss, 1980). The correction factors are discussed in Appendix 8.1. A table of temperature and cloud readings for each survey is presented in Appendix 8.3.

APPENDIX 8.3

Variables affecting aerial survey of kangaroos on Kinchega National Park, on sheep stations surrounding the park (OUTSIDE) and on Tandou sheep station. T = degrees Centigrade and C = octals of cloud cover.

YR	MTH	DAY	KINCHEGA		OUTSIDE		TANDOU	
			T	C	T	C	T	C
1973	7	1	2	0	6	0	–	–
		2	5	0			–	–
	10	1	16	0	20	0	–	–
		2	18	0			–	–
1974	3	1	21	1	26	1	–	–
		2	22	1			–	–
	6	1	8	0	3	0	–	–
		2	8	0			–	–
	9	1	8	0	6	2	–	–
		2	8	0				
1975	1	1	12	0	13	1	–	–
		2	16	0			–	–
	4	1	14	0	13	0	–	–
		2	14	0			–	–
	7	1	3	0	7	3	–	–
		2	6	1			–	–
	10	1	10	1	13	0	–	–
		2	9	0				
1976	2	1	25	4	21	0	–	–
		2	22	6			–	–
	5	1	7	1	4	2	–	–
		2	5	0			–	–
	8	1	1	0	1	0	–	–
		2	5	0			–	–
	12	1	12	0	14	0	–	–
		2	14	0				
1977	4	1	13	1	10	6	–	–
		2	10	3			–	–
	7	1	1	0	1	0	–	–
		2	1	0			–	–
	9	1	5	0	11	0	–	–
		2	5	0			–	–

YR	MTH	DAY	KINCHEGA T	C	OUTSIDE T	C	TANDOU T	C
1978	1	1	17	0	18	6	–	–
		2	14	0			–	–
	5	1	14	0	15	0	–	–
		2	12	0			–	–
	8	1	11	5	10	0	20	7
		2	4	0			–	–
	11	1	29	1	30	0	20	7
		2	19	1			–	–
1979	2	1	18	0	21		23	0
		2	18	0			12	0
	8	1	6	1	4		14	0
		2	4	1			–	–
	11	1	12	0	24		14	0
		2	12	0			–	–
1980	3	1	20	0	24	6	20	0
		2	20	0			17	0
	7	1	8	6	10	8	8	0
		2	6	1			2	0
	10	1	18	0	12	0	9	0
		2	12	0			13	0
	12	1	16	7	20	0	9	0
		2	10	1			15	0
1981	3	1	14	1	18	0	22	1
		2	20	0			18	0
	6	1	12	8	12	5	12	8
		2	7	8			7	6
	9	1	8	0	21	0	14	0
		2	20	2			17	0
	12	1	19	0	22	6	13	0
		2	9	0			20	0
1982	3	1	20	3	18	0	18	0
		2	22	3			23	7
	6	1	8	8	2	0	0	8
		2	5	0			7	0
	9	1	13	0	10	6	12	8
		2	16	0			10	2
	12	1	28	7	24	2	22	1
		2	25	7			19	0
1983	3	1	15	0	25	8	16	7
		2	14	7			16	8

YR	MTH	DAY	KINCHEGA		OUTSIDE		TANDOU	
			T	C	T	C	T	C
	6	1	12	0	8	8	11	8
		2	10	1			13	7
	9	1	21	7	10	2	16	0
		2	16	1			8	8
	12	1	25	2	30	2	25	6
		2	20	0			25	0
1984	3	1	11	0	10	0	12	0
		2	10	0			13	1
	7	1	5	2	6	6	11	7
		2	8	3			8	2

REFERENCES

Allden, W.G. & Whittaker, I.A.M. (1970). The determinants of herbage intake by grazing sheep: the interrelationship of factors influencing herbage intake and availability. *Australian Journal of Agricultural Research*, **21**, 755-66.

Anderson, D.J. (1982). The home range: a new nonparametric estimation technique. *Ecology*, **63**, 103-12.

Andrew, M.H., Noble, I.R. & Lange, R.T. (1979). A non-destructive method for estimating the weight of forage on shrubs. *Australian Rangelands Journal*, **1**, 225-31.

Andrewartha, H.G. & Birch, L.C. (1954). *The Distribution and Abundance of Animals*. Chicago: The University of Chicage Press.

Anon. (1984). *Kangaroo Management Programmes of the Australian States*. Canberra: Commonwealth of Australia.

Arnold, G.W. (1964). Factors within plant associations affecting the behaviour and performance of grazing animals. In *Grazing in Terrestrial and Marine Environments*, ed. D.J. Crisp, pp. 133-54. Oxford: Blackwell Scientific Publications.

Arnold, G.W. (1975). Herbage intake and grazing behaviour in ewes of four breeds at different physiological states. *Australian Journal of Agricultural Research*, **26**, 1017-24.

Arnold, G.W. & Dudzinski, M.L. (1967a). Studies on the diet of the grazing animal. II. The effect of physiological status in ewes and pasture availability on herbage intake. *Australian Journal of Agricultural Research*, **18**, 349-59.

Arnold, G.W. & Dudzinski, M.L. (1967b). Studies on the diet of the grazing animal. III. The effect of pasture species and pasture structure on the herbage intake of sheep. *Australian Journal of Agricultural Research*, **18**, 657-66.

Arundel, J.H., Beveridge, I. & Presidente, P.J. (1979). Parasites and pathological findings in enclosed and free-ranging populations of *Macropus rufus* (Desmarest) (Marsupialia) at Menindee, New South Wales. *Australian Wildlife Research*, **6**, 361-79.

Australian Senate. (1984). *Select Committee on Animal Welfare*, Official Hansard Report, page 1872 and pp. S2068-75.

Australian Senate. (1985). *Select Committee on Animal Welfare*, Official Hansard Report, pp. S3024-49.

Bailey, P.T. (1967). *The Ecology of the Red Kangaroo*, Megaleia rufa *(Desmarest), in North-western New South Wales*. M.Sc. Thesis, University of Sydney, Sydney.

Bailey, P.T. (1971). The red kangaroo, *Megaleia rufa* (Desmarest), in north-western New South Wales. I. Movements. *CSIRO Wildlife Research*, **16**, 11-28.

Bailey, P.T., Martensz, P.N. & Barker, R. (1971). The red kangaroo, *Megaleia rufa* (Demarest), in north-western New South Wales. II. Food. *CSIRO Wildlife Research*, **16**, 29-39.

Barker, R. D. (1986 a). An investigation into the accuracy of herbivore diet analysis. *Aust. Wildl. Res.* **13**, 559-68.

Barker, R. D. (1986 *b*). A technique to simplify herbivore diet analysis. *Aust. Wildl. Res.* **13**, 569-73.

Barker, S. (1972). *Effects of sheep stocking on the population structure of arid shrublands in South Australia.* Ph.D. Thesis. University of Adelaide, Adelaide.

Barker, S. (1979). Shrub population dynamics under grazing — within paddock studies. In *Studies of the Australian Arid Zone. IV. Chenopod Shrublands*, ed. R.D. Graetz & K.M.W. Howes, pp. 83-106. Melbourne: CSIRO.

Bayliss, P. (1980). *Kangaroos, Plants and Weather in the Semi-arid.* M.Sc. Thesis, University of Sydney, Sydney.

Bayliss, P. (1985a). The population dynamics of red and western grey kangaroos in arid New South Wales, Australia. I. Population trends and rainfall. *Journal of Animal Ecology*, **54**, 111-25.

Bayliss, P. (1985b). The population dynamics of red and western grey kangaroos in arid New South Wales, Australia. II. The numerical response function. *Journal of Animal Ecology*, **54**, 127-35.

Bayliss, P. & Giles, J. (1985). Factors affecting the visibility of kangaroos counted during aerial surveys. *Journal of Wildlife Management*, **49**, 686-92.

Beadle, N.C.W. (1948). *The Vegetation and Pastures of Western New South Wales.* Sydney: N.S.W. Government Printer.

Beatley, J.C. (1967). Survival of winter annuals in the northern Mojave desert. *Ecology*, **48**, 745-50.

Beatley, J.C. (1969). Biomass of desert winter annual plant populations in southern Nevada. *Oikos*, **20**, 261-73.

Beddington, J.R. & May, R.M. (1977). Harvesting natural populations in a randomly fluctuating environment. *Science*, **197**, 463-5.

Bell, H.M. (1973). The ecology of three macropod marsupial species in an area of open forest and savannah woodland in north Queensland. *Mammalia*, **37**, 527-44.

Bell, R.H.V. (1981). An outline of a management plan for Kasungu National Park, Malawi. In *Problems in Management of Locally Abundant Wild Mammals*, ed. P.A. Jewell, S. Holt & D. Hart, pp. 69-89. New York: Academic Press.

Bell, R.H.V. (1983). Decision-making in wildlife management with reference to problems of overpopulation. In *Management of Large Mammals in African Conservation Areas*, ed. R.N. Owen-Smith, pp. 145-72. Pretoria: HAUM Educational Publishers.

Birch, L.C. (1948). The intrinsic rate of natural increase of an insect population. *Journal of Animal Ecology*, **17**, 15-26.

Birch, L.C. (1960). The genetic factor in population ecology. *American Naturalist*, **94**, 5-24.

Black, J.L. & Kenney, P.A. (1984). Factors affecting diet selection by sheep. II. Height and density of pasture. *Australian Journal of Agricultural Research*, **35**, 565-78.

Blaisdell, J.P. (1958). Seasonal development and yield of native plants on the upper Snake River and their relation to certain climatic factors. *United States Department of Agriculture Technical Bulletin*, No. 1190.

Bosch, O.J.H. & Dudzinski, M.L. (1984). Defoliation and its effects on *Enneapogon avenaceus* and *Cenchrus ciliaris* populations during two summer growths periods in central Australian rangelands. *Australian Rangelands Journal*, **6**, 17-25.

Bowler, J.A. & Thorne, A.G. (1976). Human remains from Lake Mungo: Discovery and excavation of Lake Mungo III. In *The Origin of the Australians*, ed. R.L. Kirk & A.G. Thorne, pp. 127-38. Canberra: Australian Institute of Aboriginal Studies.

Briese, D.T. & Macauley, B.J. (1981). Food collection within an ant community in semi-arid Australia, with special reference to seed. *Australian Journal of Ecology*, **6**, 1-19.

Brooks, P.M. & Macdonald, I.A.W. (1983). The Hluhuwe-Umfolozi Reserve: An ecological case history. In *Management of Large Mammals in African Conservation Areas*, ed. R.N. Owen-Smith, pp. 51-77. Pretoria: HAUM Educational Publishers.

Brown, D. (1954). Methods of surveying and measuring vegetation. *Bulletin of the Commonwealth Bureau of Pastures and Field Crops*, No. 42.

Brown, G.H., Turner, H.N. & Dolling, C.H.S. (1968). Vital statistics for an experimental flock of merino sheep. V. The effects of ram maternal handicap and year of measurement on 10 wool and body characteristics for unselected rams. *Australian Journal of Agricultural Research*, **19**, 825-35.

Burt, W.H. (1943). Territoriality and home range concepts as applied to mammals. *Journal of Mammalogy*, **24**, 346-52.

Butlin, N.G.. (1962). Distribution of the sheep population: preliminary statistical picture, 1860-1957. In *The Simple Fleece — Studies in the Australian Wool Industry*, ed. J.A. Barnard, pp. 281-307. Melbourne: Melbourne University Press.

Calaby, J. (1966). Mammals of the upper Richmond and Clarence Rivers, New South Wales. *Technical Paper Division of Wildlife Research CSIRO*, No. 10.

Caughley, G. (1962). *The comparative ecology of the red and grey kangaroo.* M.Sc. Thesis, University of Sydney, Sydney.

Caughley, G.J. (1964). Density and dispersion of two species of kangaroo in relation to habitat. *Australian Journal of Zoology*, **12**, 238-49.

Caughley, G. (1970). Eruption of ungulate populations, with special emphasis on Himalayan thar in New Zealand. *Ecology*, **51**, 53-72.

Caughley, G. (1976a). Plant-herbivore systems. In *Theoretical Ecology: Principles and Applications*, ed. R.M. May, pp. 94-113. London: Blackwell Scientific Publications.

Caughley, G. (1976b). Wildlife management and the dynamics of ungulate populations. In *Applied Biology*, vol. 1, ed. T.H. Coaker, pp. 183-246. London: Academic Press.

Caughley, G. (1977). *Analysis of Vertebrate Populations.* London: Wiley & Sons.

Caughley, G. (1979). What is this thing called carrying capacity? In *North American Elk: Ecology, Behavior and Management*, ed. M.S. Boyce & L.D. Hayden-Wing, pp. 2-8. Laramie: University of Wyoming.

Caughley, G. (1981). What we do not know about the dynamics of large mammals. In *Dynamics of Large Mammal Populations*, ed. C.W. Fowler & T.D. Smith, pp. 361-72. New York: Wiley & Sons.

Caughley, G. (1982). Vegetation complexity and the dynamics of modelled grazing systems. *Oecologia*, **54**, 309-12.

Caughley, G. (1983). Dynamics of large animals and their relevance to culling. In *Management of Large Mammals in African Conservation Areas*, ed. R.N. Owen-Smith, pp. 115-26. Pretoria: HAUM Educational Publishers.

Caughley, G. & Birch, L.C. (1971). Rate of increase. *Journal of Wildlife Management*, **35**, 658-63.

Caughley, G., Grigg, G.C. & Short, J. (1983). How many kangaroos? *Search*, **14**, 151-2.

Caughley, G., Grigg, G.C. & Smith, L. (1985). The effect of drought on kangaroo populations. *Journal of Wildlife Management*, **49**, 679-85.

Caughley, G. & Krebs, C.J. (1983). Are big mammals simply little mammals writ large? *Oecologia*, **59**, 7-17.

Caughley, G., Sinclair, R. & Scott-Kemmis, D. (1976). Experiments in aerial survey. *Journal of Wildlife Management*, **40**, 290-300.

Caughley, J. & Bayliss, P.G. (undated [1982]). Kangaroo populations in inland New South Wales. In *Parks and Wildlife: Kangaroos and Other Macropods of New South Wales*, ed. C. Haigh, pp. 46-9. Sydney: New South Wales National Parks and Wildlife Service.

Caughley, J., Bayliss, P. & Giles, J. (1984). Trends in kangaroo numbers in western New South Wales and their relation to rainfall. *Australian Wildlife Research*, **11**, 415-22.

Chacon, E. & Stobbs, T.H. (1976). Influence of progressive defoliation of a grass sward on the eating behaviour of cattle. *Australian Journal of Agricultural Research*, **27**, 709-27.

Cheal, D.C. (1984). Hattah/Kulkyne National Park. Kangaroo numbers and habitat change. *National Parks Service of Victoria. Internal Report*. 8pp.

Chippendale, G.M. (1962). Botanical examination of kangaroo stomach contents and cattle rumen contents. *Australian Journal of Science*, **25**, 21-2.

Chippendale, G.M. (1964). Determination of plant species grazed by open range cattle in Central Australia. *Proceedings of the Australian Society of Animal Production*, **5**, 256-7.

Chippendale, G.M. (1968a). A study of the diet of cattle in central Australia as determined by rumen samples. *Primary Industries Branch, Northern Territory Administration, Technical Bulletin No. 1*. 31pp.

Chippendale, G.M. (1968b). The plants grazed by red kangaroos, *Megaleia rufa* (Demarest), in central Australia. *Proceedings of the Linnean Society of New South Wales*, **93**, 98-110.

Clark, C.W. (1976). *Mathematical Bioeconomics: The Optimum Management of Renewable Resources*. New York: Wiley - Interscience.

Cogger, H.G. (1984). Reptiles in the Australian arid zone. In *Arid Australia*, ed. H.G. Cogger & E. Cameron, pp. 235-52. Chipping Norton: The Australian Museum.

Cohen, D. (1966). Optimising reproduction in a randomly varying environment. *Journal of Theoretical Biology*, **12**, 119-29.

Condon, R. (1982). Pastoralism. In *What Future for Australia's Arid Lands?*, ed. J. Messer & G. Mosely, pp. 54-60. Victoria: Australian Conservation Foundation.

Connell, J.H. (1983). On the prevalence and relative importance of interspecific competition: evidence from field experiments. *American Naturalist*, **122**, 661-96.

Cooke, B.D. (1974). *Food and Other Resources of the Wild Rabbit*, Oryctolagus cuniculus *(L.)*. Ph.D. Thesis, University of Adelaide, Adelaide.

Cooke, B.D. (1982). Reduction of food intake and other physiological responses to a restriction of drinking water in captive wild rabbits, *Oryctolagus cuniculus* (L.). *Australian Wildlife Research*, **9**, 247-52.

Crawley, M.J. (1975). The numerical responses of insect predators to changes in prey density. *Journal of Animal Ecology*, **44**, 877-92.

Crisp, M.D. (1978). Demography and survival under grazing of three Australian semi-desert shrubs. *Oikos*, **30**, 520-8.

Crisp, M.D. & Lange, R.T. (1976). Age structure, distribution and survival under grazing of the arid-zone shrub, *Acacia burkittii*. *Oikos*, **27**, 86-92.

Croft, D.B. (1980). Radio tracking arid zone kangaroos in northwestern New South Wales. *Bulletin of Australian Mammalogy Society*, **6**(2), 30.

Croze, H., Hillman, A.K.K. & Lang, E.M. (1981). Elephants and their habitats: how do they tolerate each other. In *Dynamics of Large Mammal Populations*, ed. C.W. Fowler & T.D. Smith, pp. 297-316. New York: Wiley & Sons.

Cunningham, G.M., Mulham, W., Milthorpe, P.L. & Leigh, J.H. (1981). *Plants of Western New South Wales*. Sydney: N.S.W. Government Printer.

Curtis, B.V. (1973). The dietary preferences of the kangaroo. *Mid-Murray Field Naturalists Trust*, 6th. Report. pp. 6-7.

Davidson, J.L. & Donald, C.M. (1958). The growth of swards of subterranean clover with particular reference to leaf area. *Australian Journal of Agricultural Research*, **9**, 53-72.

Dawson, T.J., Denny, M.J.S., Russell, E.M. & Ellis, B.A. (1975). Water usage and diet preferences of free ranging kangaroos, sheep and feral goats in the Australian arid zone during summer. *Journal of Zoology, London*, **177**, 1-23.

Dawson, T.J. & Ellis, B.A. (1979). Comparison of the diets of yellow-footed rock-wallabies and sympatric herbivores in western New South Wales. *Australian Wildlife Research*, **6**, 245-54.

de Calesta, D.S., Nagy, J.G. & Bailey, J.A. (1975). Starving and refeeding mule deer. *Journal of Wildlife Management*, **39**, 663-9.

de Calesta, D.S., Nagy, J.G. & Bailey, J.A. (1977). Experiments on starvation and recovery of mule deer does. *Journal of Wildlife Management*, **41**, 81-6.

Degabriele, R. & Dawson, T.J. (1979). Metabolism and heat balance in an arboreal marsupial, the koala (*Phascolarctos cinereus*). *Journal of Comparative Physiology*, **134**, 293-302.

Denny, M.J.S. (1982). Adaptions of the red kangaroo and euro (Macropodidae) to aridity. In *Evolution of the Flora and Fauna of Arid Australia*, ed. W.R. Barker & P.J.M. Greenslade, pp. 179-83. Frewville, South Australia: Peacock Publications.

Denny, M. (undated, [1982]). Kangaroos: an historical perspective. In *Parks and Wildlife: Kangaroos and Other Macropods of New South Wales*, ed. C. Haigh, pp. 36-45. Sydney: New South Wales National Parks and Wildlife Service.

Denny, M. (1983). Animals - native and feral. In *What Future for Australia's Arid Lands?*, ed. J. Messer & G. Mosley. Victoria: Australian Conservation Foundation.

Dixon, S. (1892). The effects of settlement and pastoral occupation upon the indigenous vegetation. *Transactions of the Royal Society of South Australia*, **15**, 195-206.

Ealey, E.H.M. (1967a). Ecology of the euro, *Macropus robustus* (Gould) in north-western Australia. I. The environment and changes in euro and sheep populations. *CSIRO Wildlife Research*, **12**, 9-25.

Ealey, E.H.M. (1967b). Ecology of the euro, *Macropus robustus* (Gould) in north-western Australia. II. Behaviour, movements and drinking patterns. *CSIRO Wildlife Research*, **12**, 27-51.

Ealey, E.H.M. & Main, A.R. (1967). Ecology of the euro, *Macropus robustus* (Gould), in north-western Australia. III. Seasonal changes in nutrition. *CSIRO Wildlife Research*, **12**, 53-65.

Ellis, B.A., Russell, E.M., Dawson, T.J. & Harrop, C.J.F. (1977). Seasonal changes in diet preferences of free-ranging red kangaroos, euros and sheep in western New South Wales. *Australian Wildlife Research*, **4**, 127-44.

Eltringham, S.K. (1979). *The Ecology and Conservation of Large African Mammals*. London: Macmillan Press.

Fatchen, T.J. (1978). Change in grazed *Atriplex vesicaria* and *Kochia astrotricha* (Chenopodiaceae) populations, 1929-1974. *Transactions of the Royal Society of South Australia*, **102**, 39-42.

Fenner, F. & Myers, K. (1978). Myxoma virus and myxomatosis in retrospect: The first quarter century of a new disease. In *Viruses and Environments*, ed. E. Kurstak & K. Mavanovosch, pp. 539-70. New York: Academic Press.

Fenner, F. & Ratcliffe, F.N. (1965). *Myxomatosis*. Cambridge: Cambridge University Press.

Foley, J.C. (1957). Droughts in Australia: Review of records from earliest years of settlement to 1955. *Commonwealth of Australia Bureau of Meteorology Bulletin*, No. 43. Melbourne.

Foot, J.Z. & Romberg, B. (1965). The utilization of roughage by sheep and the red kangaroo, *Macropus rufus* (Demarest). *Australian Journal of Agricultural Research*, **16**, 429-35.

Forbes, D.K. & Tribe, D. (1970). The utilization of roughages by sheep and kangaroos. *Australian Journal of Zoology*, **18**, 247-56.

Frankel, O.H. & Soulé, M.E. (1981). *Conservation and Evolution*. Cambridge: Cambridge University Press.

Frith, H.J. (1964). Mobility of the red kangaroo, *Megaleia rufa*. *CSIRO Wildlife Research*, **9**, 1-19.

Frith, H.J. (1973). *Wildlife Conservation*. Sydney: Angus and Robertson.

Frith, H.J. & Sharman, G.B. (1964). Breeding in wild populations of the red kangaroo, *Megaleia rufa. CSIRO Wildlife Research*, **9**, 86-114.

Gaskin, D. (1982). *The Ecology of Whales and Dolphins*. London: Heinemann.

Gates, C.T. & Muirhead, W. (1967). Studies of the tolerance of *Atriplex* species. 1. Environmental characteristics and plant response of *A. vesicaria, A. nummularia* and *A. semibaccata. Australian Journal of Experimental Agriculture and Animal Husbandry*, **7**, 39-49.

Gause, G.F. (1934). *The Struggle for Existence*. Baltimore: Baltimore Press, Williams & Wilkins.

Gentilli, J. (1971). The main climatological elements. In *World Survey of Climatology*, volume 13, *Climates of Australia and New Zealand*, ed. J. Gentilli, pp. 119-88. Amsterdam: Elsevier.

Gilbert, N., Gutierrez, A.P., Frazer, B.D. & Jones, R. E. (1976). *Ecological Relationships*. Reading, Massachusetts: W.H. Freeman & Company.

Gollan, K. (1984). The Australian dingo: in the shadow of man. In *Vertebrate Zoogeography and Evolution in Australasia*, ed. M. Archer & G. Clayton, pp. 921-7. Carlisle, Western Australia: Hesperian Press.

Goodall, D.W. (1979). The chenopod shrublands of Australia - an integrated view. In *Studies of the Australian Arid Zone. IV. Chenopod Shrublands*, ed. R.D. Graetz & K.M.W. Howes, pp. 189-96. Melbourne: CSIRO.

Gooding, C.D. & Harrison, L.A. (1955). Trapping yards for kangaroos. *Journal of Agriculture of Western Australia*, **4**, 689-94.

Graetz, R.D. (1975). Biological characteristics of Australian *Acacia* and chenopodiaceous shrublands relevant to their pastoral use. In *Arid Shrublands. Proceedings of the Third Workshop of the United States/Australia Rangelands Panel*, ed. D.N. Hyder, pp. 33-9. Denver: Society of Range Management.

Graetz, R.D. & Wilson, A.D. (1980). Comparison of the diets of sheep and cattle grazing a semi-arid chenopod shrubland. *Australian Rangelands Journal*, **2**, 67-75.

Graham, A.D. (1973). *The Gardeners of Eden*. London: George Allen & Unwin.

Grant, T.R. (1973). Dominance and association among members of a captive and free-ranging group of grey kangaroos (*Macropus giganteus*). *Animal Behaviour*, **21**, 449-56.

Grant, T.R. (1974). The shapes of faecal pellets of red and grey kangaroos. *Australian Mammalogy*, **1**, 261-2.

Griffiths, M. & Barker, R. (1966). The plants eaten by sheep and by kangaroos grazing together in a paddock in south-western Queensland. *CSIRO Wildlife Research*, **11**, 145-67.

Griffiths, M., Barker, R. & MacLean, L. (1974). Further observations on the plants eaten by kangaroos and sheep grazing together in a paddock in south-western Queensland. *Australian Wildlife Research*, **1**, 27-43.

Gutterman, Y. (1981). Influence of quantity and date of rain on the dispersal and germination mechanisms, phenology, development and seed germinability in desert annual plants, and on the life cycle of geophytes and hemicryptophytes in the Negev Desert. In *Developments in Arid Zone Ecology and Environmental Quality*, ed. H. Shuval. Philadelphia: Balaban ISS.

Gwynne, M.D. & Bell, R.H.V. (1968). Selection of vegetation components by grazing ungulates in the Serengeti National Park. *Nature*, **220**, 390-3.

Hairston, N.G., Smith, F. & Slobodkin, L.B. (1960). Community structure, population control, and competition. *American Naturalist*, **94**, 421-5.

Hanks, J. (1981). Characterization of population condition. In *Dynamics of Large Mammal Populations*, ed. C.W. Fowler & T.D. Smith, pp. 47-73. New York: Wiley & Sons.

Hanks, J., Cumming, D.H., Orpen, J.L., Parry, D.F. & Warren, H.B. (1976). Growth, condition and reproduction in the impala ram (*Aepyceros melampus*). *Journal of Zoology*, **179**, 421-35.

Hanks, J., Densham, W.D., Smuts, G.L., Jooste, J.F., Joubert, S.C.J., le Roux, P. & Milstein, P. le S. (1981). Management of locally abundant mammals - the South African experience.

In *Problems in Management of Locally Abundant Wild Mammals*, ed. P.A. Jewell, S. Holt & D. Hart, pp. 21-55. New York: Academic Press.

Harper, J.L. (1977). *Population Biology of Plants*. London: Academic Press.

Harrington, G.N. (1978). The implications of goat, sheep and cattle diet to the management of Australian semi-arid woodland. In *Proceedings of First International Rangelands Congress*, pp. 447-50. Denver: Society for Range Management.

Harrington, G.N., Mills, D.M.D., Pressland, A.J. & Hodgkinson, K.C. (1984). Semi-arid woodlands. In *Management of Australia's Rangelands*, ed. G.N. Harrington, A.D. Wilson & M.D. Young, pp. 189-207. Melbourne: CSIRO.

Harrington, G.N., Oxley, R. & Tongway, D.J. (1979). The effects of European settlement and domestic livestock on the biological system in poplar box (*Eucalyptus populnea*) lands. *Australian Rangelands Journal*, **1**, 271-9.

Hassall & Associates (1982). *An Economic Study of the Western Division of New South Wales*. Volume II: Attachments. Canberra: Hassall & Associates Pty Ltd.

Haydock, K.P. & Shaw, N.H. (1975). The comparative yield method for estimating dry matter yield of pasture. *Australian Journal of Experimental Agriculture and Animal Husbandry*, **15**, 663-70.

Hayne, D.W. (1949). Calculation of size of home range. *Journal of Mammalogy*, **30**, 1-18.

Hendricksen, R. & Minson, D.J. (1980). The food intake and grazing behaviour of cattle grazing a crop of *Lablab purpureus* cv. Rongai. *Journal of Agricultural Science, Cambridge*, **95**, 547-54.

Hintz, H.F. (1969). Equine nutrition. Comparisons of digestion coefficients obtained with cattle, sheep, rabbits and horses. *The Veterinarian*, **6**, 45-51.

Holling, C.S. (1959). The components of predation as revealed by a study of small-mammal predation of the european pine sawfly. *Canadian Entomologist*, **91**, 293-320.

Holling, C.S. (1961). Principles of insect predation. *Annual Review of Entomology*, **6**, 163-82.

Hope, J. (1978). Pleistocene mammal extinctions: the problem of Mungo and Menindee, New South Wales. *Alcheringa*, **2**, 65-82.

Hope, J. (1982). Late Cainozoic vertebrate faunas and the development of aridity in Australia. In *Evolution of the Flora and Fauna of Arid Australia*, ed. W.R. Barker & P.J.M. Greenslade, pp. 85-100. Frewville, South Australia: Peacock Publications.

Horton, D.R. (1984). Red kangaroos: the last of the Australian megafauna. In *Quaternary Extinctions: a Prehistoric Revolution*, ed. P.S. Martin & R.G. Klein, pp. 639-680. Tuscon: University of Arizona Press.

Houston, D.B. (1982). *The Northern Yellowstone Elk. Ecology and Management*. New York: Macmillan Publishing Company.

Hume, I.D. (1974). Nitrogen and sulphur retention and fibre digestion by euros, red kangaroos and sheep. *Australian Journal of Zoology*, **22**, 13-23.

Ivlev, V.S. (1961). *Experimental Ecology of the Feeding of Fishes*. New Haven: Yale University Press.

Jarman, P.J. & Taylor, R.J. (1983). Ranging of eastern grey kangaroos and wallaroos on a New England pastoral property. *Australian Wildlife Research*, **10**, 33-8.

Jewell, P.A., Holt, S. & Hart, D. ed. (1981). *Problems in Management of Locally Abundant Wild Mammals*. New York: Academic Press.

Johnson, C.N. & Bayliss, P.G. (1981). Habitat selection by sex, age and reproductive class in the red kangaroo, *Macropus rufus*, in western New South Wales. *Australian Wildlife Research*, **8**, 465-74.

Johnson, C.N. & Jarman, P.J. (1983). Geographical variation in offspring sex ratios in kangaroos. *Search*, **14**, 152-4.

Jones, R.M. & Hargreaves, J.N.G. (1979). Improvements in the dry-weight-rank method for measuring botanical composition. *Grass and Forage Science*, **34**, 181-9.

Kirkpatrick, T.H. (1965). Studies of Macropodidae in Queensland. 1. Food preferences of the grey kangaroo, *Macropus major* (Shaw). *Queensland Journal of Agriculture and Animal Science*, **22**, 89-93.

Kirkpatrick, T.H. (1967). The grey kangaroo in Queensland. *Queensland Agricultural Journal*, **93**, 550-2.

Kirkpatrick, T.H. & McEvoy, J.S. (1966). Studies of Macropodidae in Queensland. 5. Effects of drought on reproduction in the grey kangaroo (*Macropus giganteus*). *Queensland Journal of Agricultural and Animal Science*, **23**, 439-42.

Klein, D.R. (1968). The introduction, increase, and crash of reindeer on St. Matthew Island. *Journal of Wildlife Management*, **32**, 350-67.

Krebs, C.J. (1971). Genetic and behavioural studies on fluctuating vole populations. In *Proceedings of the Advanced Study Institute on Dynamics of Numbers in Populations*, ed. P.J. den Boer & G.R. Gradwell, pp. 243-56. Oosterbeck.

Krefft, G. (1866). On the vertebrated animals of the Lower Murray and Darling, their habits, economy and geographical distribution. *Transactions of the Philosophical Society of New South Wales (1862-5)*, 1-33.

Krefft, G. (1871). *The Mammals of Australia*. Sydney: Government Printer.

Kuhn, T.S. (1970). *The Structure of Scientific Revolutions*. Chicago: University of Chicago Press.

Lange, R.T. & Purdie, R. (1976). Western myall (*Acacia sowdenii*), its survival prospects and management needs. *Australian Rangelands Journal*, **1**, 64-9.

Lange, R.T. & Willcocks, M.C. (1980). Experiments on the capacity of present sheep flocks to extinguish some tree populations of the South Australian arid zone. *Journal of Arid Environments*, **3**, 223-9.

Laws, R.M. (1981). Large mammal feeding strategies and related overabundance problems. In *Problems in Management of Locally Abundant Wild Mammals*, ed. P.A. Jewell, S. Holt & D. Hart, pp. 217-32. New York: Academic Press.

Lay, B. (1979). Shrub population dynamics under grazing — a long term study. In *Studies of the Australian Arid Zone. IV. Chenopod Shrublands*, ed. R.D Graetz & K.M.W. Howes, pp. 107-24. Melbourne: CSIRO.

Leigh, J.H. & Mulham, W. (1966a). Selection of diet by sheep grazing semi-arid pastures on the Riverine plain. 1. A bladder saltbush (*Atriplex vesicaria*) - cotton bush (*Kochia aphylla*) community. *Australian Journal of Experimental Agriculture and Animal Husbandry*, **6**, 460-7.

Leigh, J.H. & Mulham, W. (1966b). Selection of diet by sheep grazing semi-arid pastures on the Riverine plain. 2. A cotton bush (*Kochia aphylla*) - grassland (*Stipa variabilis - Danthonia caespitosa*) community. *Australian Journal of Experimental Agriculture and Animal Husbandry*, **6**, 468-74.

Leigh, J.H. & Mulham, W. (1967). Selection of diet by sheep grazing semi-arid pastures on the Riverine plain. 3. A bladder saltbush (*Atriplex vesicaria*) - pigface (*Disphyma australe*) community. *Australian Journal of Experimental Agriculture and Animal Husbandry*, **7**, 421-5.

Leigh, J.H., Wilson, A.D. & Mulham, W. (1968). A study of merino sheep grazing a cotton bush (*Kochia aphylla*) - grassland (*Stipa variabilis - Danthonia caespitosa*) community on the Riverine plain. *Australian Journal of Agricultural Research*, **19**, 947-61.

Lendon, C. & Ross, M.A. (1978). Vegetation. In *The Physical and Biological Features of Kunoth Paddock in Central Australia*, ed. W.A. Low, pp. 66-90. *CSIRO Division of Land Resource Management, Technical Paper*, No. 4.

Leslie, P.H. (1966). The intrinsic rate of increase and overlap of successive generations in a population of guillemots (*Uria aalge* Pont.). *Journal of Animal Ecology*, **35**, 291-301.

Lieth, H. (1973). Primary production: terrestrial ecosystems. *Human Ecology*, **1**, 303-32.

Livanes, T. (1971). Kangaroos as a resource. *Australian Zoologist*, **16**, 68-72.

Lorimer, M.S. (1978). Forage selection studies. 1. The botanical composition of forage selected by sheep grazing *Astrebla* spp. pasture in north-west Queensland. *Tropical Grasslands*, **12**, 97-108.

Low, B.S., Birk, E., Lendon, C. & Low, W.A. (1973). Community organization by cattle and kangaroos in mulga near Alice Springs, N.T. *Tropical Grasslands*, **7**, 149-56.

Low, B.S. & Low, W.A. (1975). Feeding interactions of red kangaroos and cattle in an arid ecosystem. *Proceedings Third World Conference on Animal Production*, pp. 87-94. Melbourne: University of Sydney Press.

Low, W.A., Muller, W.J., Dudzinski, M.L. & Low, B.S. (1981). Population fluctuations and range community preference of red kangaroos in central Australia. *Journal of Applied Ecology*, **18**, 27-36.

MacArthur, R.H. (1957). On the relative abundance of bird species. *Proceedings of the National Academy of Sciences USA*, **43**, 293-5.

MacArthur, R.H. (1960). On the relative abundance of species. *American Naturalist*, **94**, 25-36.

MacArthur, R.H. & Wilson, E.O. (1963). An equilibrium theory of insular zoogeography. *Evolution*, **17**, 373-87.

MacArthur, R.H. & Wilson, E.O. (1967). *The Theory of Island Biogeography*. Princeton: Princeton University Press.

McCullough, D.R. (1979). *The George Reserve Deer Herd*. Ann Arbor: Michigan University Press.

McFarlane, J.D. (1971). Exports of kangaroos. *Australian Zoologist*, **16**, 62-4.

McIntosh, D.L. (1966). The digestibility of two roughages and the rates of passage of their residues by the red kangaroo, *Megaleia rufa* (Desmarest), and the Merino sheep. *CSIRO Wildlife Research*, **11**, 125-35.

McNaughton, S.J. (1979). Grassland-herbivore dynamics. In *Serengeti — Dynamics of an Ecosystem*, ed. A.R.E. Sinclair & M. Norton-Griffiths, pp. 46-81. Chicago: University of Chicago Press.

Mannetje, L. 't & Haydock, K.P. (1963). The dry-weight-rank method for the botanical analysis of pasture. *Journal of the British Grasslands Society*, **18**, 268-75.

Marshall, W.G. (1973). Fossil vertebrate faunas from the Lake Victoria region, south western New South Wales, Australia. *Memoirs of the National Museum of Victoria*, **34**, 177-82.

Martin, P.S. (1973). The discovery of America. *Science*, **179**, 969-74.

May, R.M. (1973). *Stability and Complexity in Model Ecosystems*. New Jersey: Princeton University Press.

May, R.M. (1975). *Stability and Complexity in Model Ecosystems*. Second edition, Princeton: Princeton University Press.

May, R.M. (1976). Harvesting whale and fish populations. *Nature*, **263**, 91-2.

May, R.M., Beddington, J.R., Horwood, J.W. & Shepherd, J.G. (1978). Exploiting natural populations in an uncertain world. *Mathematical Biosciences*, **42**, 219-52.

Mentis, M.T. (1977). Stocking rates and carrying capacities for ungulates on African rangelands. *South African Journal of Wildlife Research*, **7**, 89-98.

Mentis, M.T. & Duke, R.R. (1976). Carrying capacities of natural veld in Natal for large wild herbivores. *South African Journal of Wildlife Research*, **6**, 65-74.

Merrilees, D. (1973). Fossiliferous deposits at Lake Tandou, New South Wales, Australia. *Memoirs of the National Museum of Victoria*, **34**, 151-71.

Mitchell, T.L. (1839). *Three Expeditions into the Interior of Australia; with Descriptions of the Recently Explored Region of Australia Felix, and of the Present Colony of New South Wales*, 2nd edn. London: T. & W. Boone. [Facsimile edition, Adelaide, Libraries Board of South Australia, 1965].

Moore, C.W. (1953a). The vegetation of the south-eastern Riverina, New South Wales. I. The climax communities. *Australian Journal of Botany*, **1**, 485-547.

Moore, C.W. (1953b). The vegetation of the south-eastern Riverina, New South Wales. II. The disclimax communities. *Australian Journal of Botany*, **1**, 548-67.

Moore, R.M. (1959). Ecological observation of plant communities grazed by sheep in Australia. *Monographie Biologicae*, **8**, 500-13.

Moore, R.M. (1962). Effects of the sheep industry on Australian vegetation. In *The Simple Fleece. Studies in the Australian Wool Industry*, ed. J.A. Barnard, pp. 170-83. Melbourne: Melbourne University Press.

Moore, R.M. (1975). Australian arid shrublands. In *Arid Shrublands. Proceedings of the Third Workshop of the United States/Australia Rangelands Panel*, ed. D.N. Hyder, pp. 6-11. Denver: Society of Range Management.

Moore, R.M. & Perry, R.A. (1970). Vegetation. In *Australian Grasslands*, ed. R.M. Moore, pp. 59-73. Canberra: Australian National University Press.

Morris, R.F., Cheshire, W.F., Miller, C.A. & Mott, D.G. (1958). The numerical response of avian and mammalian predators during a gradation of the spruce budworm. *Ecology*, **39**, 487-94.

Mott, J.J. (1972). Germination studies on some annual species from an arid region of Western Australia. *Journal of Ecology*, **60**, 293-304.

Mott, J.J. & McComb, A.J. (1975). The role of photoperiod and temperature in controlling the phenology of three annual species from an arid region of Western Australia. *Journal of Ecology*, **63**, 633-41.

Newman, J.C. & Condon, R.W. (1969). Land use and present condition. In *Arid Lands of Australia*, ed. R.O. Slayter & R.A. Perry, pp. 105-32. Canberra: Australian National University Press.

Newsome, A.E. (1964). Anoestrus in the red kangaroo, *Megaleia rufa* (Desmarest). *Australian Journal of Zoology*, **12**, 9-17.

Newsome, A.E. (1965a). The abundance of red kangaroos, *Megaleia rufa* (Desmarest), in central Australia. *Australian Journal of Zoology*, **13**, 269-87.

Newsome, A.E. (1965b). The distribution of red kangaroos, *Megaleia rufa* (Desmarest), about sources of persistent food and water in central Australia. *Australian Journal of Zoology*, **13**, 289-99.

Newsome, A.E. (1965c). Reproduction in natural populations of the red kangaroo, *Megaleia rufa* (Desmarest), in central Australia. *Australian Journal of Zoology*, **13**, 735-59.

Newsome, A.E. (1966). The influence of food on breeding in the red kangaroo in central Australia. *CSIRO Wildlife Research*, **11**, 187-96.

Newsome, A.E. (1971). Competition between wildlife and domestic livestock. *The Australian Veterinary Journal*, **47**, 577-86.

Newsome, A.E. (1975). An ecological comparison of the two arid-zone kangaroos of Australia and their anomalous prosperity since the introduction of ruminant stock to their enviroment. *The Quarterly Review of Biology*, **50**, 389-424.

Newsome, A.E. (1977). Imbalance in the sex ratio and age structure of the red kangaroo, *Macropus rufus*, in central Australia. In *The Biology of Marsupials*, ed. B. Stonehouse & D. Gilmour, pp. 221-33. London: Macmillan.

New South Wales Parliament. (1983). *Joint Select Committee of the Legislative Council and Legislative Assembly to Enquire into the Western Division of New South Wales*. Sydney: D.West, Government Printer.

Noble, I.R. (1975). *Computer simulations of sheep grazing in the arid zone*. Ph.D Thesis, University of Adelaide, Adelaide.

Noble, I.R. (1977). Long-term biomass dynamics in an arid chenopod shrub community at Koonamore, South Australia. *Australian Journal of Botany*, **25**, 639-53.

Noy-Meir, I. (1973). Desert ecosystems: environments and producers. *Annual Review of Ecology and Systematics*, **4**, 25-51.

Noy-Meir, I. (1975). Stability of grazing systems: an application of predator-prey graphs. *Journal of Ecology*, **63**, 459-81.

Noy-Meir, I. (1978). *Stability and Complexity in Model Ecosystems*. New Jersey: Princeton University Press.

Noy-Meir, I. (1979). Stability in simple grazing models: effects of explicit functions. *Journal of Theoretical Biology*, **71**, 347-80.

Orr, D.M. (1978). *Effects of grazing Astrebla grassland in central western Queensland*. M.Sc. Thesis, University of Queensland, Brisbane.

Orr, D.M. (1981). Changes in the quantitative floristics in some *Astrebla* spp. (Mitchell grass) communities in south-western Queensland in relation to trends in seasonal rainfall. *Australian Journal of Botany*, **29**, 533-45.

Owen-Smith, R.N. ed. (1983a). *Management of Large Mammals in African Conservation Areas*. Pretoria: HAUM Educational Publishers.

Owen-Smith, R.N. (1983b). Dispersal and the dynamics of large herbivores in enclosed areas: implications for management. In *Management of Large Mammals in African Conservation Areas*, ed. R.N. Owen-Smith, pp. 127-43. Pretoria: HAUM Educational Publishers.

Oxley, J. (1820). *Journals of Two Expeditions into the Interior of New South Wales*. London: John Murray. [Facsimile edition, Adelaide, Libraries Board of South Australia, 1964].

Oxley, R. (1979). The perennial chenopod pasture lands of Australia. In *Studies of the Australian Arid Zone. IV. Chenopod Shrublands*, ed. R.D. Graetz & K.M.W. Howes, pp. 1-4. Melbourne: CSIRO.

Parra, R. (1978). Comparison of foregut and hindgut fermentation in herbivores. In *The Ecology of Arboreal Folivores*, ed. G.G. Montgomery, pp. 205-29. Washington: Smithsonian Institute Press.

Pearson, L.C. (1965a). Primary production in grazed and ungrazed desert communities of eastern Idaho. *Ecology*, **46**, 278-85.

Pearson, L.C. (1965b). Primary productivity in a northern desert area. *Oikos*, **15**, 211-28.

Perry, R.A. (1977). Rangeland management for livestock production in semi-arid and arid Australia. In *The Impact of Herbivores on Arid and Semi-Arid Rangelands, Proceedings of the Second United States/Australia Rangelands Panel, Adelaide, 1972*, pp. 311-16. Perth, Western Australia: Australian Rangelands Society.

Pettigrew, C. & Lyons, M. (1979). Royal National Park — A history. *Parks and Wildlife*, **2** (3-4), 15-30.

Philander, S.G.H. (1983). El Nino Southern Oscillation phenomena. *Nature, London*, **302**, 295-301.

Pianka, E.R. (1969). Habitat specificity, speciation and species density in Australian desert lizards. *Ecology*, **50**, 498-502.

Pienaar, U. de V. (1983). Management by intervention: the pragmatic/economic option. In *Management of Large Mammals in African Conservation Areas*, ed. R.N. Owen-Smith, pp. 23-36. Pretoria: HAUM Educational Publishers.

Pieper, R.D. (1978). *Measurement techniques for herbaceous and shrubby vegetation*. Las Cruces: Mimeographic Services, New Mexico State University.

Pilton, P. (1961). Reproduction in the great grey kangaroo. *Nature, London*, **189**, 984-5.

Poole, W.E. (1973). A study in breeding in grey kangaroos, *Macropus giganteus* Shaw and *M. fuliginosus* (Desmarest), in central New South Wales. *Australian Journal of Zoology*, **21**, 183-212.

Poole, W.E. (1975). Reproduction in the two species of grey kangaroos, *Macropus giganteus* (Shaw) and *M. fuliginosus* (Desmarest). II. Gestation, parturition and pouch life. *Australian Journal of Zoology*, **23**, 333-53.

Poole, W.E. (1978). Management of kangaroo harvesting in Australia. Occasional Paper No. 2. Canberra: Australian National Parks and Wildlife Service.

Poole, W.E. & Catling, P.C. (1974). Reproduction in the two species of grey kangaroos, *Macropus giganteus* Shaw and *M. fuliginosus* (Desmarest). I. Sexual maturity and oestrus. *Australian Journal of Zoology*, **22**, 277-302.

Porter, R.N. (1977). Wildlife management objectives and practices for the Hluhluwe Game Reserve and the northern corridor. Unpublished, Natal Parks Board, 9pp. [Not seen. Quoted by Brooks & Macdonald (1983).]

Priddel, D. (1983). *The Movements of Red and Western Grey Kangaroos in Western New South Wales*. Ph.D. Thesis, University of Sydney, Sydney.

Prince, R.I.T. (1976). *Comparative studies of aspects of nutritional and related physiology in macropod marsupials*. Ph.D. Thesis, University of Western Australia, Perth.

Prince, R.I.T. (1984a). Exploitation of kangaroos and wallabies in Western Australia. I. A review to 1970 with special emphasis on the red and western grey kangaroos. *Wildlife Research Bulletin of Western Australia*, **13**, 1-78.

Prince, R.I.T. (1984b). Exploitation of kangaroos and wallabies in Western Australia. II. Exploitation and management of the red kangaroo: 1970-1979. *Wildlife Research Bulletin of Western Australia*, **14**, 1-145.

Riney, T. (1955). Evaluating condition of free-ranging red deer (*Cervus elaphus*) with special reference to New Zealand. *New Zealand Journal of Science and Technology, Section B*, **36**, 429-63.

Riney, T. (1964). *The Impact of Introductions of Large Herbivores on the Tropical Environment*. I.U.C.N. Publications, New Series, **4**, 261-73.

Robards, G. (1978). Regional and seasonal variation in wool growth throughout Australia. In *Physiological & Environmental Limitations to Wool Growth*, ed. J.L. Black & P.J. Reis, pp. 1-42. Armidale: University of New England Press.

Robards, G., Leigh, J.H. & Mulham, W.E. (1967). Selection of diet by sheep grazing semi-arid pastures on the Riverine plain. 4. A grassland (*Danthonia caespitosa*) community. *Australian Journal of Experimental Agriculture and Animal Husbandry*, **7**, 426-33.

Robertson, G.G. (1986). The mortality of kangaroos in drought. *Australian Wildlife Research*, **13**, 349-54.

Robertson, G.G. & Gepp, B. (1982). Capture of kangaroos by 'stunning'. *Australian Wildlife Research*, **9**, 393-6.

Rosenweig, M. (1971). Paradox of enrichment: destabilisation of exploitation ecosystems in ecological time. *Science*, **171**, 385-7.

Ross, M.A. (1969). An integrated approach to the ecology of arid Australia. *Proceedings of the Ecological Society of Australia*, **4**, 67-81.

Ross, M.A. (1976). The effects of temperature on germination and early growth of three plant species indigenous to Central Australia. *Australian Journal of Ecology*, **1**, 259-63.

Ross, M.A. & Lendon, C. (1973). Productivity of *Eragrostis eriopoda* in a mulga community. *Tropical Grasslands*, **7**, 111-16.

Russell, E.M. (1974). The biology of kangaroos (Marsupialia - Macropodidae). *Mammal Review*, **4**, 1-59.

Sadleir, R.M. (1965). Reproduction in two species of kangaroo (*Macropus robustus* and *Megaleia rufa*) in the arid Pilbara region of Western Australia. *Proceedings of the Zoological Society of London*, **145**, 239-61.

Schoener, T.W. (1983). Field experiments on interspecific competition. *American Naturalist*, **122**, 240-85.

Schoener, T.W. (1985). Some comments on Connell's and my reviews of field experiments on interspecific competition. *American Naturalist*, **125**, 730-40.

Schullery, P. (1984). *Mountain Time*. New York: Nick Lyons Books.

Scott-Kemmis, D. (1979). *The Distribution and Abundance of Grey Kangaroos in Relation to Environment in Arid and Semi-arid Areas of New South Wales*. M.Sc. Thesis, University of Sydney, Sydney.

Sharma, M.L., Tunny, J. & Tongway, D.J. (1972). Seasonal changes in sodium and chloride concentration of saltbush (*Atriplex* spp.) leaves as related to soil and plant water potential. *Australian Journal of Agricultural Research*, **23**, 1007-19.

Sharman, G.B. & Calaby, J.H. (1964). Reproductive behaviour in the red kangaroo, *Megaleia rufa*, in captivity. *CSIRO Wildlife Research*, **9**, 58-85.

Sharman, G.B., Frith, H.J. & Calaby, J.H. (1964). Growth of pouch young, tooth eruption, and age determination in the red kangaroo, *Megaleia rufa. CSIRO Wildlife Research*, **9**, 20-49.

Sharon, D. (1972a). The spottiness of rainfall in a desert area. *Journal of Hydrology*, **17**, 161-75.

Sharon, D. (1972b). Spatial analysis of rainfall data from dense networks. *Bulletin of the International Association of Hydrological Science*, **17**, 291-300.

Sharon, D. (1979). Correlation analysis of the Jordan Valley rainfall field. *Monthly Weather Review*, **107**, 1042-7.

Sharrow, S.H. & Motazedian, I. (1983). A comparison of three methods for estimating forage disappearance. *Journal of Range Management*, **36**, 469-71.

Shepherd, N.C. (1981). *Protection of coastal waterbirds*. Law Hons. Thesis, Macquarie University, Sydney.

Shepherd, N.C. (1983). The feasibility of farming kangaroos. *Australian Rangelands Journal*, **5**, 35-44.

Shepherd, N.C. & Giles, J.R. (undated [1982]). Kangaroo management in New South Wales. In *Parks and Wildlife: Kangaroos and Other Macropods of New South Wales*, ed. C. Haigh, pp. 50-5. Sydney: New South Wales National Parks and Wildlife Service.

Short, J. (1985). The functional response of kangaroos, sheep and rabbits in an arid grazing system. *Journal of Applied Ecology*, **22**, 435-47.

Short, J. (1986). The effect of pasture availability on food intake, species selection and grazing behaviour of kangaroos. *Journal of Applied Ecology*, **23**, 559-71.

Short, J. & Bayliss, P. (1985). Bias in aerial survey estimates of kangaroos. *Journal of Applied Ecology*, **22**, 415-22.

Silcock, R.G. (1977). A study of the fate of seedlings growing on sandy red earths in the Charleville district, Queensland. *Australian Journal of Botany*, **25**, 337-46.

Silcock, R.G. (1980). Seedling characteristics of tropical pasture species and their implications for ease of establishment. *Tropical Grasslands*, **14**, 174-80.

Sinclair, A.R.E. (1975). The resource limitation of trophic levels in tropical grassland ecosystems. *Journal of Animal Ecology*, **44**, 497-520.

Sinclair, A.R.E. (1977). *The African Buffalo. A Study of Resource Limitation of Populations*. Chicago: The University of Chicago Press.

Sinclair, A.R.E. & Duncan, P. (1972). Indices of condition in tropical ruminants. *East African Wildlife Journal*, **10**, 143-9.

Singh, G., Kershaw, A.P. & Clark, R. (1981). Quaternary vegetation and fire history in Australia. In *Fire and the Australian Biota*, ed. A.M. Gill, R.H. Groves & I.R. Noble, pp. 23-54. Canberra: Australian Academy of Science.

Slobodkin, L.B., Smith, F.E. & Hairston, N.G. (1967). Regulation in terrestrial ecosystems, and the implied balance of nature. *American Naturalist*, **101**, 109-24.

Solomon, M.E. (1949). The natural control of animal populations. *Journal of Animal Ecology*, **18**, 1-35.

Southwood, T.R.E. (1966). *Ecological Methods*. London: Meuthuen.

Specht, R.L. (1973). Vegetation. In *The Australian Environment*, ed. G.W. Leeper, pp. 44-67. Melbourne: Melbourne University Press.

Squires, V.R. (1980). Chemical and botanical composition of the diets of oesophageally fistulated sheep, cattle and goats in a semi-arid *Eucalyptus populnea* woodland community in north-west New South Wales. *Australian Rangelands Journal*, **2**, 94-103.

Squires, V.R. (1981). Dietary overlap between sheep, cattle and goats when grazing in common. *Journal of Range Mangement*, **35**, 116-19.

Squires, V.R. (1982). Competitive interactions in the dietary preference of kangaroos and sheep, cattle and goats in inland Australia. *Journal of Arid Environments*, **5**, 337-45.

Stoddart, L.A., Smith, A.D. & Box, T.W. (1975). *Range Management*. New York: McGraw-Hill Book Company.

Storr, G.M. (1968). Diet of kangaroos (*Megaleia rufa* and *Macropus robustus*) and merino sheep near Port Hedland, Western Australia. *Journal of the Royal Society of Western Australia*, **51**, 25-32.

Strahan, R. (1983). *The Australian Museum Complete Book of Australian Mammals*. Sydney: Angus & Robertson.

Streten, N.A. (1981). Southern hemisphere sea surface temperature variability and apparent associations with Australian rainfall. *Journal of Geophysical Research*, **86**, 485-97.

Streten, N.A. (1983). Extreme distributions of Australian annual rainfall in relation to sea surface temperature. *Journal of Climatology*, **3**, 143-53.

Strugnell, R.G. & Pigott, C.D. (1978). Biomass, shoot production and grazing of two grasslands in the Rwenzori National Park, Uganda. *Journal of Ecology*, **66**, 73-96.

Sturt, C. (1849). *Narrative of an Expedition into Central Australia Performed Under the Authority of Her Majesty's Government, During the Years 1844, 5 and 6, Together with a Notice of the Province of South Australia in 1847*. London: T. & W. Boone. [Facsimile edition, Adelaide, Libraries Board of South Australia, 1965.]

Suijdendorp, H. (1955). Changes in pastoral vegetation can provide a guide to management. *Journal of Agriculture of Western Australia*, **4**, 683-7.

Tanner, J.T. (1975). The stability and intrinsic growth rates of prey and predator populations. *Ecology*, **56**, 855-67.

Tedford, R.H. (1967). The fossil macropodidae from Lake Menindee, New South Wales. *University of California Publications on Geological Science*, **64**, 1-165.

Trevis, L.T. (1958). Germination and growth of ephemerals induced by sprinkling a sand-desert. *Ecology*, **39**, 681-8.

Trumble, H.C. (1935). The relation of pasture development to environmental factors in South Australia. *Agricultural Journal of South Australia*, **38**, 1460-87.

Tyndale-Biscoe, H. (1973). *Life of Marsupials*. London: Edward Arnold.

Van der Kley, F.K. (1956). A simple method for the accurate estimation of daily variations in the quality and quantity of herbage consumed by rotationally grazed cattle and sheep. *Netherlands Journal of Agricultural Science*, **4**, 197-204.

Van Soest, P.J. (1973). Collaborative study of acid-detergent fibre and lignin. *Journal of the Association of Official Analytical Chemists*, **56**, 781-4.

Van Soest, P.J. & Wine, R.H. (1968). Use of detergents in the analysis of fibrous feeds. IV Determination of plant cell-wall constituents. *Journal of the Association of Official Analytical Chemists*, **50**, 50-5.

Van Vuren, D. & Coblentz, B.E. (1985). Kidney weight variation and the kidney fat index: and evaluation. *Journal of Wildlife Management*, **49**, 177-9.

Wakefield, N.A. (1966). Mammals of the Blandowski Expedition to north-west Victoria, 1856-57. *Proceedings of the Royal Society of Victoria*, **79**, 371-91.

Walker, B.H. (1981). Stability properties of semi-arid savannas in southern African game reserves. In *Problems in management of Locally Abundant Wild Mammals*, ed. P.A. Jewell, S. Holt & D. Hart, pp. 57-67. New York: Academic Press.

Walker, B.H. & Goodman, P.S. (1983). Some implications of ecosystem properties for wildlife management. In *Management of Large Mammals in African Conservation Areas*, ed. R.N. Owen-Smith, pp. 79-91. Pretoria: HAUM Educational Publishers.

Walter, H. (1955). Le facteur eau dans les régions arides et sa signification pour l'organisation de la végétation dans les contrées sub-tropicales. *Collogue sur les régions écologiques du globe*, **59**, 27-39.

Wells, R.T. (1973). *Physiological and behavioral adaptions of the hairy-nosed wombat (Lasiorhinus latifrons Owen) to its arid environment*. Ph.D. Thesis, University of Adelaide, Adelaide.

Westoby, M. (1979-80). Elements of a theory of vegetation dynamics in arid rangelands. *Israeli Journal of Botany*, **28**, 169-94.

Wheeler, J.L., Reardon, T.F. & Lambourne, L.J. (1963). The effect of pasture availability and shearing stress on herbage intake of grazing sheep. *Australian Journal of Agricultural Research*, **14**, 364-72.

White, P.J. & O'Connell, J.F. (1982). *A Prehistory of Australia, New Guinea and Sahul*. Sydney: Academic Press.

Williams, O.B. (1961). Studies in the ecology of the Riverine Plain. III. Phenology of a *Danthonia caespitosa* Gaudich. grassland. *Australian Journal of Agricultural Research*, **12**, 247-59.

Williams, O.B. (1968). That uneasy state between animal and plant in the manipulated situation. *Proceedings of the Ecological Society of Australia*, **3**, 167-74.

Williams, O.B. (1974). Vegetation improvement and grazing management. In *Studies of the Australian Arid Zone. II. Animal Production*, ed. A.D. Wilson, pp. 127-43. Melbourne: CSIRO.

Williams, O.B. & Oxley, R.E. (1979). Historical aspects of the chenopod shrublands. In *Studies of the Australian Arid Zone. IV Chenopod Shrublands*, ed. R.D. Graetz & K.M.W. Howes, pp. 5-16. Melbourne: CSIRO.

Williams, O.B. & Roe, R. (1975). Management of arid grasslands for sheep: plant demography of six grasses in relation to climate and grazing. *Proceedings of the Ecological Society of Australia*, **9**, 142-56.

Willoughby, W.M. (1959). Limitations to animal production imposed by seasonal fluctuations in pasture and by management procedures. *Australian Journal of Agricultural Research*, **10**, 248-68.

Wilson, A.D. (1974). Nutrition of sheep and cattle in Australian arid areas. In *Studies of the Australian Arid Zone. II. Animal Production*, ed. A.D. Wilson, pp. 74-84. Melbourne: CSIRO.

Wilson, A.D. (1976). Composition of sheep and cattle grazing on a semi-arid grassland. *Australian Journal of Agricultural Research*, **27**, 155-62.

Wilson, A.D. (1977). The digestibility and voluntary intake of the leaves of trees and shrubs by sheep and goats. *Australian Journal of Agricultural Research*, **28**, 501-8.

Wilson, A.D. & Graetz, R.D. (1979). Management of semi-arid and arid rangelends of Australia. In *Management of semi-arid ecosystems*, ed. B.H. Walker, pp. 83-111. Amsterdam: Elsevier Scientific Publishing Company.

Wilson, A.D., Harrington, G.N. & Beale, I.F. (1984). Grazing management. In *Management of Australia's Rangelands*, ed. G.N. Harrington, A.D. Wilson & M.D. Young, pp. 129-39. Melbourne: CSIRO.

Wilson, A.D., Leigh, J.H., Hindley, N.L. & Mulham, W.E. (1975). Comparison of the diets of goats and sheep on a *Casuarina cristata - Heterodendrum oleifolium* woodland community in western New South Wales. *Australian Journal of Experimental Agriculture and Animal Husbandry*, **15**, 45-53.

Wilson, A.D., Leigh, J.H. & Mulham, W.E. (1969). A study of merino sheep grazing a bladder saltbush (*Atriplex vesicaria*) - cotton-bush (*Kochia aphylla*), community on the Riverine plain. *Australian Journal of Agricultural Research*, **20**, 1123-36.

Wilson, G.R. (1975). Age structure of populations of kangaroos (Macropodidae) taken by professional shooters in New South Wales. *Australian Wildlife Research*, **2**, 1-9.

Wood, D.H. (1980). The demography of a rabbit population in an arid region of New South Wales, Australia, *Journal of Animal Ecology*, **49**, 55-79.

Young, J.A., Evans, R.A. & Major, J. (1972). Alien plants in the Great Basin. *Journal of Range Management*, **25**, 194-201.

Zar, J.H. (1974). *Biostatistical Analysis*. Englewood Cliffs, N.J.: Prentice-Hall Incorporated.

AUTHOR INDEX

SUBJECT INDEX

Subject index

253